PERVERTED TRUTH EXPOSED

How Progressive philosophy has corrupted Science
First in the Modern Mythology Series

**WHAT DO WE REALLY KNOW ABOUT
"LIFE, THE UNIVERSE AND EVERYTHING"?**
(Hitchhikers Guide to the Galaxy
by Douglas Adams, 1979)

T. K. KISER

PERVERTED TRUTH EXPOSED

World Ahead Press is a division of WND Books. The views and opinions expressed in this book are those of the author and do not necessarily reflect the official policy or position or WND Books.

Paperback ISBN: 978-1-944212-18-6
eBook ISBN: 978-1-944212-19-3

Printed in the United States of America
16 17 18 19 20 21 LSI 9 8 7 6 5 4 3 2 1

CONTENTS

INTRODUCTION

Much of what we think we know about the universe and life is really **belief, based on faith.** The main purpose in writing this book is to show that much of what we think we know about science is actually not science at all but philosophy. While technology is certainly real, and observation and experimental data are real, inferences, conclusions, projections, deduction from pure reason of what (we believe) must be, models and other facts, especially about events in the past, are really beliefs disguised as facts. Many of the "facts" that fall into this category are either irrelevant or detrimental to true science, and are only relevant to supporting certain social and cultural paradigms.[1]

Through all of this philosophy disguised as science is a common thread of progressivism. The universe, the Earth, life, everything are assumed (on faith) to be naturally progressing from simpler through more organized and complex toward perfection. Progressivism is teleological[2] because progress is assumed to have a purpose – to move toward perfection. As such, it requires faith. It takes as much faith to believe in a universe and life that invented themselves, as it takes to believe in a creator, whether it is one that set up the initial conditions and walked away, or one that continues to orchestrate its functions or cares about mankind on a personal basis.

I want to present the wonderful complexity of the universe and life itself, in order to:

> It ain't so much the things we don't know that get us into trouble.
>
> It's the things we know that ain't so.
>
> *- Artemus Ward aka Charles Farrar Browne*

- teach you to use critical thinking about what you may read or hear

- logically examine some of today's most cherished scientific theories

- present some possible alternatives for some of the poorly supported ones

- present evidences that science and a belief in God are not incompatible

In other words, I am seeking to get you to think and to open the stage for rational discussions, not emotional diatribes in these areas. **The goal is truth.** In so doing, I expect to anger and upset some in both the theological and scientific communities. My apologies to both.

Behind all other philosophies, there are two basic ways of looking at the universe, the materialistic and the transcendental views.

NOTE: Do not confuse transcendental as used here with the so-called Transcendentalism movement of the nineteenth century which was really a Naturalistic or agnostic philosophy. Henry David Thoreau is a good example of that philosophy. I call him "the man who mistook nature for God."

NOTE: Materialism, in the philosophical sense, is not the love of things as the term is used today, having been corrupted by the popular culture. It is really the denial of the existence of anything beyond the material world.

The materialist view says the physical universe that we see and interact with is all that there is, and it is its own explanation for being. The transcendent view says there must be something more behind and above it all, an overarching force that is responsible for existence itself and the physical laws that give order to the universe. The opinions above reflect my transcendental beliefs that there is something beyond the material that is beyond the reach of science. As a transcendentalist and a scientist, I have no reservations about science and religion being compatible.

Since materialists only believe in what they can see and touch, pure materialists are necessarily atheists and have a great deal of reservations about religion, especially as it relates to science. Even the agnostic, who is uncertain about a transcendent quality to the universe, is not comfortable with the notion that science and religion are compatible, since he believes that there is no way for us to know about anything outside the material world.

> Try and penetrate with our limited means the secrets of nature and you will find that, behind all the discernible concatenations, there remains something subtle, intangible and inexplicable.
>
> *- Albert Einstein*

Joke: What do you get when you cross an insomniac agnostic and a dyslexic?

A person who stays up all night wondering if there is a Dog.

Chico Marx (often attributed to Groucho Marx)

Science is the pursuit of truth about the predictable, repeatable and measurable aspects of the universe with which we can or could conceivably interact[3].

Anything beyond that is not science but philosophy, no matter how much mathematics or "supporting" data is attached. A one-time event that cannot be tested or repeated, such as the origin of the universe or the origin of life cannot be elucidated purely through science.

Materialists/atheists sometimes use the fact that we can only observe and test the physical world as proof that there is nothing else beyond that. Although science, in its truest sense, is an unbiased quest for the truth, that does not mean the people in the sciences are always unbiased or have no hidden agendas outside of science. Similarly, although religion seeks answers from a God centered perspective, man's interpretations of God's revealed truth may at times lead to error through misunderstandings or cultural bias.

As a Christian, I believe God's revelations in the Bible are true, but I also believe that man's interpretations can sometimes be wrong. An example of this from history is the supposed flat-Earth belief in the European Middle Ages, (which, by the way, is a nineteenth century myth[4]). Those few holding this belief had interpreted "the four corners of the earth," in Isaiah[5] and Revelation[6] to mean the Earth had four literal corners and was thus flat. What was obviously meant were the four directions. Even at that time, the belief in a flat Earth was not widespread and a spherical Earth had been common knowledge[7] especially for sailors who clearly saw the curvature of the horizon and saw sails appear over it before the ships appeared. It was evident in earlier ages that the Earth cast a circular shadow on the moon, especially during lunar eclipses.

The ancient Greeks and other cultures in antiquity knew the Earth was a sphere and actually calculated its circumference by triangulation. You may find it hard to believe, but when I was a child, I went to school with the children of an uneducated family that believed in a flat Earth with four corners. It was a real crisis of faith for them when man first went into space and sent back pictures of the obviously spherical Earth and then later actually went to the moon.

> The discovery of truth is prevented more effectively, not by the false appearance things present and which mislead into error, not directly by weakness of the reasoning powers, but by preconceived opinion, by prejudice.
>
> *- Arthur Schopenhauer*

Contrary to what you may have been taught, true science and Christianity are thoroughly compatible, and Christians have contributed greatly to the foundations of science. Anyone who believes in God must also believe that God invented science and gave us the ability to understand the laws of the universe. More than any other religion, Christianity taught that God had created an orderly, understandable universe that obeys natural laws. In this view, it is left to individuals to discover what that order and those laws are.

- Science seeks the truth of WHAT, WHERE, WHEN and HOW things happen.

- Religion seeks the truth of WHY things happen, WHO might have caused them and for "what" ultimate Purpose.

Therefore, we would not expect science to say anything at all about the why and who of religion, or for religion to necessarily seek spiritual answers in the what, where and how of science. The two are separate but complementary parts of the whole picture. They both seek to understand our "world" but approach it from different directions. Some agenda driven scientists who campaign against a belief in God have stepped beyond science into what amounts to a religion of evangelical atheism, which is also based on faith. Just because science can only study the material world does not mean there is nothing else outside it that is beyond the reach of science.

A closed minded, dogmatic scientist is an oxymoron[8] and is not a true scientist at all regardless of her credentials. Science is a developing discipline where new knowledge sometimes radically changes accepted theories. The universe is filled with mysteries and unanswered or unanswerable questions. A true scientist must keep an open mind about all possibilities and admit that some things are not known or not even knowable. It is not necessary to make up clever stories to compensate for the missing knowledge, much less teach such speculative beliefs as settled science. It is OK to say "I don't know" unless you are pushing an agenda other than truth. When scientists fill in gaps in knowledge with clever "just-so" stories, it is a Science of the Gaps[9] and is not science at all.

As a scientist, I use the tools of science to search for truth about our world. I think it is safe to say that there are prejudices and fiercely held beliefs on both sides. But beliefs are not facts no matter how many are convinced that they are true. It makes no sense to hold on to a view that has been shown to be in error, simply because it is "accepted and established." Nor is it useful to stoop to name calling, slurs and other put downs of those on the other side of an issue. Consensus is alien to science and is

> To be a faultless member of the flock, first one must be a sheep.
>
> - Albert Einstein

only properly used for opinions, not facts, in cultural and political settings. Consensus can lead into egregious errors. Remember, before Copernicus and Galileo the consensus was that the entire universe revolved around the Earth. Truth is never the result of a popular vote.

Nowhere is it more evident today than in the subjects of evolution and the origins of life and the universe. Both sides of these questions are often so steeped in their own emotional capital that their judgment and willingness to honestly debate may be clouded. For some, there are no grounds for discussion, much less any compromises. The other side is often viewed as either evil, ignorant or having sinister hidden motives. Science by definition is the quest for truth about our world; it has no business being closed minded and dogmatic to the point of preaching against religious belief or God based on materialist beliefs and prejudices.

The actions of a scientist are closer to a religion than true science when he insists that religion is bad and that everyone must believe as he does. Unfortunately, evangelical atheism has often used science to further its "religion." This has also been true of the Progressive and Socialist political agendas that often go hand in hand with it. Both philosophies have effectively used the all too gullible press as a weapon in their war on both religion and truth. It is impossible to separate science, politics, popular culture, religion and the role of the press in the search for truth. They are all irreversibly entangled.

Because evolution is such an emotionally charged subject, when I started to write this book I was inclined to give it a minor role compared to other mysteries. However, when I studied it further from all sides, I realized that evolution was the beginning of and the role model for promoting most of the later dogma disguised as science, and belief disguised as truth. Its presentation in the popular press rather than peer reviewed scientific journals[10], the political style of rhetoric used to promote it, and the types of defensive arguments employed, taught later generations how to promote their points of view in other fields. In many areas, science has been hijacked by materialists who promote progressive and anti-God views through philosophical story-telling disguised as science.

Evolution is also a good example of entrenched dogma that the adherents insist must be accepted without question. As such, it is closer to myth than science. Science is never chiseled in stone. It is a changing discipline where nothing and no one are held sacred or inviolable. New knowledge is always to be encouraged and should never be seen as a threat. If any area of science does stifle dissenting views, it is not science any more; it is a religion or at least a strongly held philosophy. Unfortunately, the fields of evolution, origin of life, cosmology and particle physics have become so dogmatic that dissent or differing views are attacked, defunded and blocked from publication in scientific journals.

This book hopes to look at many areas where science and religion have clashed in the past and where they continue to disagree. We will explore some of the wonderful workings of our universe from the largest astronomical scale to the tiniest subatomic particles, and at life itself while challenging some of the most cherished scientific beliefs. At times you will be asked to lay aside your prejudices and follow a logical exploration in a quest for truth.

NOTE: Since this book is intended to be read and understood by people at all levels of education and experience, and since my purpose is to communicate rather than to impress the reader, I have tried to keep it as simple as possible in both language and content. I have not used technical terms, jargon or other five-dollar words except as needed and always with clarification, definition, and other explanations. When a complex subject is discussed, I have attempted to give simple, readable background information so that anyone can understand it or at least grasp the fundamentals. I hope I have been successful in this and apologize if I have failed to adequately explain anything well enough. For those who do not need such explanations, you can just skip over the endnotes, explanatory boxes, et cetera, and read through the text directly. The explanations in the endnotes are especially useful for those who are new to the subject matter.

CAUTIONS IN VIEWING OBJECTS

Beware of determining and declaring your Opinion suddenly on any Object: for Imagination often gets the start of Judgment, and makes People believe they see Things which better Observation will convince them could not possibly be seen: therefore assert nothing till after repeated Experiments and Examinations in all Lights, and in all Positions.

When you employ the Microscope shake off all Prejudice, nor harbour any favorite opinions: for, if you do, it is not unlikely Fancy will betray you into Error, and make you think you see what you would wish to see.

Remember, that Truth alone is the Matter you are in Search after; and if you have been mistaken, let no Vanity reduce you to persist in your Mistake.

Pass no Judgment upon things over-extended by Force, or contracted by Dryness, or in any Manner out of their natural State without making suitable Allowances.

MDCCLXXXV

From chapter XV of Henry Baker's *Of Microscopes and the Discoveries Made Thereby*, 1785, London

PART 1

TRUTH: SCIENCE, PHILOSOPHY AND CRITICAL THINKING

CHAPTER 1

TRUTH

What do we really know? What is fact and what is opinion? What is knowledge and what is belief, and can we know the difference? Isn't science about facts and religion about faith? Well, not entirely. Science, with all of its trappings of mathematics, still is subject to interpretation, ie, belief. There is as much faith in science as in anything else we do. Consensus and computer models do not change a belief into a fact.

DO WE KNOW:

- that there was a Big Bang that started the universe?

- that black holes, parallel universes, exotic dark matter or dark energy exist?

- how all of the elements and physical laws originated?

- how the galaxies, stars, the solar system, planets, the Earth or the moon were formed?

- the true distances to other galaxies?

- the age of the universe, our galaxy or the Earth?

- that the universe, including space itself, is expanding?

- that the fourth dimension or multiple dimensions exist?

- that a dimension known as space-time exists?

- what gravity is?

- what time is?

- what life is?

- that life spontaneously arose from a soup of chemicals?

- that all species evolved gradually from a common ancestor?

- that the mind is just a program created by the brain?

- what consciousness, thought or memory are?

- what sleep is?

- what instinct is?

- why we have free will and are not just robotic slaves to our genes?

- why we have abilities and skills that are not necessary or are detrimental to survival?

The answer to most of these and many other questions about science and our understanding of our world is MAYBE, NO, or PROBABLY NOT.

The bad news is that we don't know as much as we thought we knew.
The good news is that we don't know as much as we thought we knew.

Bringing some accepted scientific facts or the evidence supporting them into question will not tear down our knowledge base. On the contrary, it will open doors to more exciting discoveries, unconstrained by fixed paradigms[1] or established systems into which they must be fitted. By questioning everything, we can look at all things with fresh eyes and with minds open to all possibilities, regardless of established beliefs. This should lead to more scientific study and discoveries, not less. Robust scientific theories and real facts will be strengthened by such questioning.

Only the theories without proper basis or support will suffer. Even those will benefit from fresh approaches that may come closer to solving some of the remaining mysteries than is currently possible. It is to our benefit that true understanding can develop unconstrained by dogma.[2] Fixed dogma tends to constrain and inhibit new knowledge, especially if the new knowledge does not fit neatly into the established picture.

A closed-minded scientist is an oxymoron[3]. He is either a true scientist in pursuit of the truth, no matter how uncomfortable, or he is a closed-minded person practicing science. He can never be

> Michael Faraday warned against the tendency of the mind 'to rest on an assumption' and when it appears to fit in with other knowledge to forget that it has not been proved.
>
> *- W. I. B. Beveridge, The Art of Scientific Investigation*

both closed minded and a true scientist. A true scientist is never threatened by new knowledge or differing viewpoints. As a matter of fact, real research is often benefited by a devil's advocate[4] who challenges both the results and conclusions in such a way as to sharpen the scientist's perception of the problem. I have used this approach to sharpen the understanding of others in research projects with which I have been involved.

Many people, who wear the title of scientist today, including renowned professors, are not true scientists at all. They are practicing "science" with closed minds, prejudices, and agendas that often have nothing to do with the pursuit of actual truth. Their agendas can include pride and ego, career building, getting agenda-driven government grants, protecting turf, atheism and politics among others. Data that does not fit the agenda is often ignored or is reinterpreted in order to force it into their preconceived world view.

Any newly discovered facts or contrary opinions that are uncomfortable or challenge the preconceived picture are vehemently opposed by these so-called experts. Both the research and the scientist who is guilty of presenting the new facts may be attacked or ridiculed with the aim of running him out of the profession as a fraud, crack-pot, ignoramus or worse – a Christian!

> Great spirits have always encountered violent opposition from mediocre minds.
>
> - *Albert Einstein*

In a world where scientists know a lot more about their own technical fields of study than most other people do, is it really possible to question such authority? The answer is an emphatic YES. Appeal to authority is one of many logical fallacies[5] and should never be a basis for TRUTH. (But why, Mommy? Because I said so.) Nor is consensus proof of the truth of a theory. (But, Mommy, everyone else is doing it.) Consensus may have some value in social and political settings, but it has no valid place in science. Truth is never the result of a committee vote.

We will look at some of these errors in logic later as we discuss how to recognize them. The so-called superstars in science, both current and historical, are not immune to erroneous ideas and should never be put on a pedestal as infallible. The really great breakthroughs of history, such as the sun-centered solar system, have stepped outside the prevailing wisdom of the time. While most people can't know everything scientists know, we can look at specific examples within any particular field to judge whether the information supports the theory as being either:

- Truism - a self-evident truth, especially one too obvious or unimportant to mention, eg, the phases of the moon exist. This is also called a tautology.

- Fact - that which is real or demonstrated to be true, eg, the phases of the moon are caused by the direction of the sun relative to the moon and the Earth.

- Theory supported by sufficient solid evidence and experiments to believe it is probably true. A Law is a theory that has passed this test over time.

OR

- Theory based mostly on assumptions or speculations that make it fit into a nested system of beliefs.

Assumptions and beliefs, especially those based on insufficient evidence, are more akin to pseudoscience, philosophy or religion than science. There is a term for that. It's called metaphysics,[6] meaning beyond the physical, and it lies outside of true science and in the realm of philosophy. **Science is only concerned with predictable, repeatable, and measurable aspects of the universe with which we can or could conceivably interact.** To be scientific, a theory must also be testable, verifiable, and falsifiable.

One-time events such as the origin of the universe or life are necessarily outside of science. For example, even if we ultimately can experimentally design processes that may have led to the origin of life, it does not necessarily verify the sequence of events that actually occurred, much less verify that it occurred by random chance based on progressive beliefs. We cannot know these things. They lie in the realm of the unknowable. The connections that exist are based, at best, on forensics[7] and at worst on speculations. Because we can never go back in time to verify their assumptions, the fields of cosmology, evolution, and origin of life studies, have often reverted to clever stories or computer models as a substitute for experimental evidence.

Correlations and similarities may be assumed, rightly or wrongly, to be causes or antecedents. Descriptions of **what** they believe happened are often substituted for explanations of **how** they actually happened or even **that** they actually happened and thus explain nothing. The absence of a known alternative explanation (story) doesn't make the accepted one true. While these stories may be based on a set of known facts, much of the story is based either on interpretation or projection, in order to fit it into the accepted theory. It may also be purely scientific speculation or just plain guessing.

Scientific speculation in itself is an oxymoron. If it is science it is not speculative; if it is speculation it is not fully science, no matter how much evidence there is. A speculation based on observations, called a hypothesis, should be the beginning point of scientific inquiry, not the end product. These stories, in spite of their scientific jargon and mathematical trappings, have a lot in common with the ancient myths, creation and otherwise, which all cultures have and which are meant to be accepted without question.

Truth is defined as the body of real things, events and facts. It is the actual state of affairs rather than that which is manifest or assumed. "Manifest" means that which appears evident, but is not necessarily true. Many things are more apparent than real. For example, the sun moving across the sky is manifest, but the Earth actually rotates under the sun. Astronomers talk about apparent brightness of distant stars or galaxies to distinguish between their actual brightness and the reduced brightness caused by their distance.

> There is something fascinating about science. One gets such wholesale returns of conjecture out of such a trifling investment of facts.
>
> *- Mark Twain, Life on the Mississippi*

What is true and how can we recognize it? What is a theory and how does it differ from fact? How does science move from hypothesis (statement) to theory (accepted explanation or belief) and then to fact (truth)? How can we know what we know? How can we recognize the difference between truth and mere speculation that is based on assumptions, but that lacks sufficient relevant evidence to be considered a proven fact?

Science is heuristic[8]; it starts with an unproven hypothesis (statement of assumptions) based on initial specific observations; only then does science set out to do experiments to prove or disprove the hypothesis in a general sense. Experiments are designed to both support and refute the hypothesis. Finding mere examples that appear to fit do not qualify as experimental results. They are often referred to as anecdotal[9] and shouldn't carry much scientific weight. The purpose of experiments is not merely to supply examples that support a hypothesis, but to determine if there is any evidence that would disprove it. Experiments must be designed to challenge as well as support the hypothesis. If verifiable experimental results don't fit the hypothesis, then the hypothesis must be rejected or modified to fit the facts revealed in the experimental data.

When enough experimental evidence supports it and none has disproved it, at some point the hypothesis is accepted as a theory. A theory is a hypothesis with enough evidence to believe it may be true, but which lacks absolute proof to be considered a fact. When something is shown to be true by repeated and varied experiments over time, at some point it becomes accepted as a Law. That does not necessarily guarantee that it is true, only that it is more likely to be true than other explanations. This is the simple picture, but it is often more complicated than that.

Science starts with 1) assumptions, 2) defines the rules for examining the observed phenomenon, 3) decides which data to collect or keep and which to ignore, and then 4) interprets the results. The theory may be thought to be true but, in reality, may be false if any of these four stages is not done correctly or is done with bias. Even a theory that is accepted as Law may be erroneous if the basic assumptions on which it is built are incorrect or even slightly off the mark.

A classic example of this is the Ptolemaic Earth-centered universe of Aristotle. The European assumption was that the entire universe revolves around the Earth. The evidence was the movement of sun, moon, stars, and planets across the sky and a failure to sense any movement of the Earth. Some of the planets moved in rather strange loops (epicycles) in which they were seen to move against the fixed stars both forward and reversed in nearly circular loops. This was simply interpreted as a curious fact and was incorporated as a part of the Earth-centered picture.

> All truth passes through three stages. First, it is ridiculed. Second, it is violently opposed. Third, it is accepted as being self-evident.
>
> - Arthur Schopenhauer

The Earth-centered view was not changed until Copernicus' theory (at first an untested hypothesis) and Galileo's observations (experiments) provided evidence, but not proof, refuting the Earth-centered universe theory in favor of a sun-centered planetary system. The Earth's daily rotation and its revolution in its yearly orbit around the sun caused all of the observations. With a sun-centered system, the epicycles were recognized as the apparent movement of slower moving outer planets against the stars caused by movement of the Earth in its yearly orbit and were therefore no longer needed.

As a side note, although a sun centered theory had been proposed by the Greeks as early as 400 BC, Ptolemy and Aristotle rejected this picture based on the aesthetic assumption that orbits would be perfectly circular, not elliptical. Some of the resistance of Copernicus' theory, which also maintained circularity, was based on the fact that his calculations did not exactly match reality. The epicycles of the Earth-centered view approximated the elliptical orbits better than the circular orbits of the Copernican theory. Galileo also maintained a belief in circular orbits. Johannes Kepler later formulated the theory of elliptical orbits that agreed better with reality.

This is also an excellent example of a theory becoming a Law over time. Even though the later development of Newton's celestial mechanics supported it with strong evidence, absolute proof had to wait for direct observation from space. If someone had not questioned the Earth-centered view, we would have needed to wait for space flight to prove or refute it. As a matter of fact, celestial mechanics based on the Earth-centered view probably would have been derived wrong and space flight might never have happened at all. This is also a good example of a new discovery going against accepted dogma.

There are two ways of searching for the truth. One is the inductive approach which starts with specific examples and sets out to prove a general principle. The other is deduction which starts with a general statement from pure reason and then seeks to fill in the specifics to support it. In the case of evolution, origin of life studies, cosmology, and particle physics, deduction is the chosen method because the actual events happened in the past and cannot be directly tested. Deduction is weaker than induction and must be more strongly supported by experiment and observation to be considered probably true.

Most of science is true; it is the interpretation and implied or induced conclusions as well as deduction from pure reason that most often are in question. The question of whether a theory reflects an actual fact or just a belief should always be considered. Critical thinking is required to examine whether a scientific "fact" (read "theory") is true or not. It is not enough to take the word of experts without question. We must examine what it is based on, how the conclusion was formed and sometimes even the agenda it supports.

Today we can add computer models to Disraeli's list. Statistics and computer models can "prove" anything, given the right set of data, the right rules of the game and the right interpretation of the results. The little book *How to Lie with Statistics* by Darrell Huff [10] shows how easily facts can be distorted by the choice of data, the means of collection, the assumptions made and the type of mathematical analysis done.

> There are three types of lies: lies, damn lies and statistics.
>
> *- Benjamin Disraeli*
> *(Often attributed to Mark Twain)*

One example of misuse of statistics is mistaking correlation for cause and effect. When two things occur together, it doesn't mean one caused the other, any more than saying everyone drinks water and everyone dies; therefore drinking water causes death. Water can be fatal, but you usually have to inhale quite a bit of it. The environmental, medical, and social sciences are especially prone to this type of error. Are all those smoking deaths really due to smoking or is smoking an indicator of risky lifestyle choices in general or even genetics? Does poverty cause crime, as is supposed, or does stinking thinking cause most of the poverty and crime? Good question.

Statistics based on opinion polls or questions about behavior are even more unreliable. The "goodness" of the data depends on the agenda of the survey, the location, the people chosen, the wording and the order in which the questions are asked, the attitude and appearance of the surveyor, the popularity of the "right" answers, and the need for the respondent to please the surveyor or to appear intelligent or socially astute. What and who is left out of the survey is as important as what and who is included.

One classic example of a survey gone wrong is the newspaper headline Dewey Defeats Truman in the 1948 presidential election, based on a poll (survey). The poll used the telephone. Only wealthier and better-educated people had telephones at that time, so conclusions were wrong because the respondents were not representative of the voting majority.

It is always important to know who and how many people were excluded either by the design of the survey or because they chose not to participate. Many people today are suspicious of unsolicited political surveys by phone and refuse to answer. Any results of these polls only reflect those willing to participate, which may not be representative of the whole population. In order to have any validity, it is important to know what percentage of those contacted actually participated.

In surveys, it is advisable to know how the questions were asked. Was it written or oral, group or individual? Was it conducted at the mall, a church, school, business, or an election polling place? Were rural, suburban, or blighted city neighborhoods excluded from the sample? What were the agendas, attitudes, ages, and appearances of the surveyors and the respondents? The choice of words and the order of questions can also affect the answers. If the survey is put together in such a way as to bias the results in favor or against a particular point of view or cause, then the answers will reflect this. They will get the answers they want.

Example of a survey about dog ownership

Case 1: Following questions about the problems and responsibilities of dog ownership, asking, "You wouldn't want to buy such a mangy, flea-bitten hound, would you?"

Case 2: Without first setting up the premise that dogs are a lot of trouble and instead asking about your feelings about dogs as companions, asking, "Would you want to buy a fine dog like this?"

In the first case, a negative answer is expected; in the second case, a more positive answer is expected. Supposedly the surveys are both about dog ownership, but in reality, the choice and order of questions and their wording have biased the results.

Although surveys are supposed to randomly select a cross-section of people, there is a tendency for surveyors to unconsciously pick certain types of people and ask the questions in a way that hints at the expected answer. Many people, being eager to please the interviewer, will LIE (big surprise).

For instance, if the question is asked, "How often do you pick your nose?" Because it is commonly perceived to be a disgusting habit, most people will say, "Never" (that is unless the person is a middle school boy whose agenda is to be outrageous and shock the surveyor). Is that a true statistic? Hardly.

It gets even worse when you ask middle or high school students about their use of tobacco or alcohol or their sexual experiences. Most will lie to appear grown up or cool, based on the current definition of cool among their peers. As with the all too common boys' locker room sexual prowess stories (lies) used to appear cool and intimidate peers, surveys on such subjects produce similar misleading information.

It doesn't tell you what they really do; it only tells you what they SAY they do. Unfortunately, such responses are often reported in the press as true behavior, especially if it fits the surveyor's or reporter's social agenda. Many years ago, when I was teaching elementary school in a depressed area, it was not uncommon for boys to claim to smoke and get drunk on weekends. This was true even for those in the first and second grade. They had heard the older boys bragging about it and wanted to appear cool, too. Of course, unknown to the younger kids, many of the older boys were also exaggerating or lying for the same purpose.

One recent example was a study done that proved that students would rather suffer a painful shock than be bored with nothing to do[11]. It actually only showed the obvious, that college students are curious when presented with an opportunity to investigate (the shock) and nothing else to do. It had nothing to do with preferring shock to boredom.

In observational or experimental statistics, the goodness of the conclusions is based on the design of the experiment, the data chosen or ignored, the significance placed upon it, the assumed link between two or more sets of data, whether cause and effect is assumed, the size of the sample, and the exceptions excluded.

It is also based on the type of mathematical treatment used. What is reported, the mean (average) or the median (middle value), and was the data skewed in one direction or another? What is the spread and distribution of the data? Is there enough variability to render the conclusion meaningless? What factors and data were ignored or discarded? Was there an appropriate control sample used to verify the results? Were positive and negative interactions between factors, ie, synergy,[12] considered? Was the factor being tested one of the major causes of the effect or was it only a minor factor among several so that it is actually overshadowed by other, more important factors that were not considered? Are all factors known or understood? Was enough data collected to draw statistically significant conclusions? Is the experiment reproducible by others with the same results?

I've seen many mistakes in design of experiments and statistical analysis of data, including trying to do statistics on three data points, which is ludicrous. One engineer I worked with conducted a week-long intensive study that proved he had solved the humidity sensitivity of a photopolymer system with his new moisture-resistant oxygen-barrier film. Everyone was very excited about his report until I produced electron photomicrographs, requested by him after the study, which showed his new barrier was not even a film, but more of an open foam. The system did not get worse with rising humidity because it was an open system that was as bad as it could get at normal humidity.

If he had looked at his film first, he would not have wasted his time on a worthless study. Needless to say, he was very embarrassed and upset with me, although I was only the messenger. I hated to do it, but the truth had to be told. Otherwise our research would have taken a wrong path. If you start with the wrong assumptions or fail to use the proper controls, you can come to incredibly erroneous conclusions. Mere background noise can be mistaken for real information.

I once worked with an industrial research scientist who could take any set of widely dispersed data points, arrayed on a graph like the stars in a constellation, and fit a trend line through them. By making certain assumptions such as forcing the line through zero, and throwing out the data that didn't fit his preconceived assumptions as "outliers," he would draw up a formula and proclaim, "It's intuitively obvious." On a few occasions, when this was particularly difficult, he would take the logarithm (log) or even the log of the log of the data to help it fit better. Never mind that if you log any set of data enough times and plot it on log scale graphs you will get a (nearly) straight line. Data manipulation like this doesn't make the conclusion real; it just fits it into the preconceived idea and agenda.

A logarithm is the exponent of a number, usually 10, written 10^2. An exponent is the number of times minus one (n-1) that a number is multiplied by itself, or the fractional equivalent thereof. The log of 100 is 2, so it is 10 multiplied by itself 1 time or 10 x 10=100. The log of 1000 is 3 or 10 x 10 x 10=1000. Now there is a lot more difference between 100 and 1000 than there is between 2 and 3.

You can see where this is heading: large differences can appear to be minor. Now if I log the log, the loglog of 100 is 0.301 and the loglog of 1000 is 0.477. It certainly appears they are very similar but in truth they are 900 units apart!

Cherry picking information is one way of supporting a favorite hypothesis while ignoring conflicting data or assigning it to experimental error or random anomalies. This is also known as science a la cutting room floor, referring to the practice of cutting and splicing movies during production (now done digitally). One way to do this is to only include those factors that produce the outcome desired or to weight them (give them greater mathematical importance) in a favorable way. It is always wise to ask what was ignored or eliminated from consideration to make the data fit the theory. As in life, in science you often get what you look for - and miss what you don't. If renewal of your government (or industry or advocacy group) funded grant depends on your research supporting the desired picture, it is likely you will do only the tests that support it.

Computer models are often used to predict outcomes, but are only as good as the initial assumptions, the data used, the weight they are given and the rules of the game employed to arrive at the predicted conclusions. If all significant factors affecting outcomes are not known or considered, that omission will bias the results. Where computer models are used to predict the future, they are especially unreliable. Weather prediction is a good example of this. Even after satellite data and standard models became available, the weather forecaster can only predict general trends, based on conditions in the immediate past and present, by assuming that all of the factors are known and that current trends will continue.

Due to unpredictable factors such as local phenomena like lakes and mountains or changing speed and path of travel, there is still a good chance that your barbeque or outdoor wedding will get rained out on a day predicted to have only clear skies, or that those predicted snow flurries will actually dump a foot of snow on you. The truth is, we understand very little about how our atmosphere creates weather on a local level, let alone on a global scale. We're getting better at it, but there is still a long way to go.

That said, how reliable might season-long hurricane predictions or global climate change models be that extend years into the future? Just like the local weather, the conclusions can only be based on our current knowledge of how the system works, conditions in the immediate past and present, the assumptions made about future trends and the factors that are assumed to affect them. If all of the important factors affecting weather patterns or climate are not included or known, then the predictions can be very unreliable.

For example, in 2006, many weather forecasters predicted a higher than normal number of Atlantic hurricanes based on the previous year's high number and certain assumptions about the rising temperature of the ocean. As it turned out, fewer than the normal number of hurricanes occurred and not one hit the continental United States. Only two made landfall after being downgraded to tropical storms. Because some important factors were not known or understood, the prediction was unreliable.

It is interesting to note that one of the original computer models to predict global warming was created by the same person who created the computer model in the 1970s that predicted an impending ice age. Dr. James Hansen of NASA's Goddard Institute of Space Studies championed both models. There are many variations on this theme, but none of the models are shown to be any better at predicting outcomes.

Computer models follow the rules of GIGO: Garbage In, Garbage Out. Conclusions may be very wrong if every single factor affecting a system is not known, or if some are unpredictable or are given the wrong weight (emphasis) in calculations. Also, in computer models, very small input errors can result in huge output errors, especially over longer periods of time, which can magnify the effects. This is true whether we are talking about the weather, global warming, plate tectonics (formerly called continental drift), species extinction scenarios, or the expansion of the universe.

The Butterfly effect, which states that the movement of a butterfly's wings can affect weather elsewhere in the world at a future time, is an illustration of the sensitivity of initial conditions in computer models. It was never meant to imply that an actual butterfly could affect the real weather elsewhere, but has repeatedly (intentionally?) been interpreted as such by environmentalists to drive their agenda. This sensitivity to initial conditions has also been adopted by other theoretical scientific fields such as chaos theory and quantum physics.

Computer models are really sets of mathematical formulas and have a lot in common with computer games. They 1) start with certain initial conditions (assumptions), 2) define the rules of play (which data is used and the way it is to be used and analyzed) and 3) produce a result based on those conditions.

Mathematics and computer models are used by cosmologists and particle physicists to "prove" all sorts of theories for which little or no experimental evidence is possible. If mathematical equations can be formulated that work for ten dimensions and that fit the prevailing assumptions, then it is assumed to be proven that there really ARE ten dimensions, and that the models or theories based on them must also be true.

Never mind that it is all a mathematical or computer construction based on the initial assumptions and rules employed. Never mind that the "fact" can never be demonstrated or tested by experiment because, among other reasons, we are not capable of perceiving more than the three dimensions we see. Remember, **science is only concerned with predictable, repeatable and measurable aspects of the universe with which we can or could conceivably interact.**

If it works out on paper (or the computer) as not being mathematically impossible, then it must be true! That's like saying that the characters in a video game, which is also a mathematical construct, are real because they exist in the game and you can make them do things by following the rules of the game. While it is true that real systems can be beautifully analyzed and defined mathematically, is the opposite true? Can mathematics, no matter how complex, be interpreted to mean that it defines real systems in the real world? Ah, there is the problem. Although something can be made to happen mathematically, that doesn't necessarily make it true. Mathematics and computer models based on mathematics can only be approximate descriptions of real systems. Unfortunately, much of cosmology and particle physics has confused the ability of mathematics to describe reality with reality itself.

Such is the world of parallel universes, multiple bubble universes, and things such as vacuum energy or quantum fluctuations that create matter out of empty space. Is any of it true? Maybe it is; maybe it's not. Without experimental evidence to back it up, we can only be certain that it has not been shown to be mathematically impossible—yet. We'll look at some of these in more detail later.

> The important thing is not to stop questioning; curiosity has its own reason for existing. One cannot help but be in awe when contemplating the mysteries of eternity, of life, of the marvelous structure of reality. It is enough if one tries merely to comprehend a little of the mystery every day. The important thing is not to stop questioning; never lose a holy curiosity.
>
> *- Albert Einstein*

CHAPTER 2

KNOWLEDGE AND BELIEF

STAGES OF KNOWLEDGE AND BELIEF

There are three main stages or levels when it comes to scientific knowledge of the universe and belief in God or any type of higher power – the Naïve, the Sophisticate, and the Sage.

- The Naïve are relatively non-scientific and embrace a belief in God without question. They see the hand of God in everything, based on wonder and awe of the world around them. They have little knowledge of the science behind it all. This is not to disparage them. The complexity of life and nature is sufficient for them to believe that there must be a God behind it all, and, to a great degree, rightly so. Nature really is awesomely complex and mind-bogglingly immense. It is at the same time both terrible and wonderful. These people may or may not believe the scientific theories presented in the popular press, but have little basis for critically analyzing them.

> Better is a poor and a wise child than an old and foolish king, who will no more be admonished.
>
> *Ecclesiastes 4:13*

- Sophisticates are the Status Quo[1] Regurgitators (SQR), who embrace the authoritative standard model or party line as gospel and inviolable. These intellectuals and scientists are so sure they have all of the answers that they see other views as inferior or just plain wrong. They believe everything can be explained by science naturally and they reject the possibility of other scientific explanations, things that are beyond the capability of science, and a creator, God, or any other type of transcendent[2] power. To them, the universe is the answer to itself. They accept on blind faith the established scientific picture without question. If new data seems at odds with the established picture, then ad hoc[3] statements must be added to bring it into line.

> The Cosmos is all that is or ever was or ever will be.
>
> *- Carl Sagan, Cosmos*

Although they are often academic leaders in their fields, they are really shallow thinkers. Their faith in science and historical or popular scientific leaders takes on a fervor rivaling any religion. As such, they fiercely defend their beliefs and feel compelled to force those beliefs on others and to suppress opposing views or questions. They fail to see the bigger picture or its implications and often focus on the parts rather than the whole. They also fail to see that their view is really a religion in itself, based on faith in prevailing scientific beliefs and heroes. Through hubris and political action, they often rise to the top of their departments, mostly because they are so sure they have superior knowledge and abilities.

- Sages are the Independent Thinkers (IT) who have the same or better knowledge of science as the sophisticates, but who don't just follow lock-step and regurgitate what they are told without question. They are willing to admit that there are some things that science has yet to explain fully, and may never be able to prove. They are willing to consider other possible explanations for observations outside the status quo. The Independent Thinkers have a deeper understanding because they are not constrained by loyalty to any dogma. They also may see beyond it to a bigger picture and its implications. A deeper understanding of the implications of *life, the universe and everything*[4] often leads them to the inescapable conclusion that there must be something beyond and outside of it all, whether it is God in the classical sense or something else equally powerful.

> Try and penetrate with our limited means the secrets of nature and you will find that, behind all the discernible concatenations, there remains something subtle, intangible and inexplicable.
>
> *- Albert Einstein*

The SQR, (Status Quo Regurgitators) who are largely in control of the scientific and intellectual academic establishment through political promotion by like-minded peers, (but by no means in the majority), tend to ridicule and suppress the other two categories. As a way of silencing scientific opposition, they often lump the other two categories together in a disparaging way, although the two are at opposite ends of the scientific understanding and belief scales. The one accepts God on faith alone; the other uses superior scientific knowledge to sometimes reach similar conclusions. The SQR tend to have shallower understanding based on knowledge that they themselves limit because of their faith in a scientific dogma that is accepted without question. This view is sometimes referred to as COWDUNG, COnventional Wisdom of the DomiNant Group[5].

The SQR also sometimes have hidden agendas that have more to do with philosophy, politics and a progressive worldview than with science or truth. As rabid defenders and promoters of the status quo, they are really more religious and evangelical[6] than either of the other two categories, although their faith is in scientific dogma and a form of materialism[7] known as naturalism[8].

Because members of the media (including the press) are impressed by media-savvy scientists and tend to be relatively unscientific or at best superficially scientific, they are the willing, though often unwitting, pawns of the SQR who never miss an opportunity to promote their views. This tendency results in press reports and documentaries that sensationalize and legitimize theories supporting the established view and that denigrate those holding other views or who question the status quo. The press, being led astray by the scientific establishment, makes sure the public believes that the regurgitators have superior knowledge and that what they say must be accepted as fact without question. Fact and proof are words loosely used by the press when theory and evidence would be the better choice of words.

Government funding of research by political bureaucracies, especially but not exclusively at universities and government agencies, also tends to support the status quo by directing funding to popular projects while effectively de-funding any research into areas contrary to COWDUNG and anything disparaged or not advocated in the popular press. Government funding decisions draw heavily on public opinion and politics, which have little or no connection to the real value of the research under consideration.

Universities rely on government grants to support research projects, so they are not likely to encourage any research not apt to get a government grant. Thus academic research and knowledge are limited by popular funding choices. The only way that academic freedom can be fostered is to remove the ability of government to actually influence the direction of research. Of course, that still leaves SQR department heads to determine those directions and deny any proposal not in line with the status quo. This also includes editors and peer review panels of scientific journals.

The public often feels unarmed to do battle with such well-armed and formidable foes. There are two extreme paths people can take. One is to accept blindly what the regurgitators tell them right down to the no God part. The other is to reject knowledge outright and hold to faith in God alone. Most people, however, take a more moderate view and accept a blending of both. Many are willing to accept at face value what they are told about how the universe works, but are not willing to concede that God had nothing to do with it. The more logical way is to question everything and draw informed conclusions based on both knowledge and faith.

Unfortunately, scientists who are Independent Thinkers will get nowhere in their academic careers if they rock the boat too much, so they tend to remain silent about any ideas that question or are at odds with the current established view. Although they may still question accepted norms and even conduct private experiments pursuing the truth, they seldom express those doubts publicly, much less attempt to publish scientific papers on them – at least more than once. That would be career suicide, leading to ridicule and expulsion from academia, branded as a pseudo-scientist, or worse.

Enough examples of such discipline exist for the message to be clear. Step out of line and you're done. A few brave souls wait until retirement to express their views, but most choose to keep their reputations unsullied. The result is not a search for excellence but a search for mediocrity and a stifling of research that is in pursuit of the truth. Only those rare scientists that have reached Nobel laureate or similar independent status dare pursue unpopular studies or views. Even then those studies are usually rejected by scientific journals because they do not fit the agenda.

> Men occasionally stumble over the truth, but most of them pick themselves up and hurry off as if nothing had happened.
>
> *- Sir Winston Churchill*

The quest for new knowledge and an understanding of life and our universe suffer from these constraints. Academia, rather than being a free forum for discussion of many different views, has become a closed system that feeds only on those ideas that support the established norm. While using the medieval church as an example of intellectual suppression, academia itself has taken on many of the characteristics of the inquisition. One glaring example of this is the denial of publication in prestigious scientific journals of peer reviewed research papers that even hint at the possibility of Intelligent Design, regardless of the scientific rigor of the research.

Opponents, eg, Darwinian evolutionists and atheists, then point to a lack of peer reviewed published papers as proof that there is no real scientific work being conducted in the area. The same is true of some environmental studies that fail to support the very popular theory of global warming or at least climate change.

Remember, a consensus does not a fact make, and contrary to claims, this subject is far from consensus anyway. Here again, COWDUNG wins out and those not playing the game by the rules are either excluded or punished.

Another phenomenon that may not be obvious to those outside the scientific community is the obligatory inclusion of phrases supporting the prevailing dogma or popular cause of the day in any published work, even when the body of the work has little or no connection to it. I have seen this in biochemistry textbooks and peer-reviewed[9] journal articles on everything from astronomy to social science, so-called. Even when the authors attempt to stay neutral, it appears that they are often forced to include at least a mention of the dogma in order to get published. This is certainly true of peer reviewed scientific articles, but popular mainstream publications are also guilty of it. Some of the most popular causes promoted today in this manner are evolution, cosmology, and climate change. Many professional and popular publications have crossed the line from support to advocacy.

One example I recently ran across in the popular press was an article in National Geographic magazine on hummingbirds.[10] It was a beautifully done picture article about the wondrous features of these remarkable birds. At one point the statement is made, "In death, their delicate hollow bones almost never fossilize," and goes on to talk about a remarkable find of a fossil in Germany that indicates they once existed outside of the Americas. Then, there is this curious statement: "'True hummingbirds', says Schuchmann, 'evolved in Brazil's eastern forests, where they competed with insects for flower nectar.'" The statement, interestingly attributed to another, not only claims a very specific place of origin, but claims to know of actual behavior. Neither of which can be known with any certainty. In an article about modern-day hummingbirds, this out-of-place statement seems to be inserted to fill the obligatory homage to evolutionary theory expected by the publisher, a strong advocate of Darwinian evolutionary theory.

Global warming, or as it has come to be called, climate change, is a subject that has recently been exposed as exemplifying all of these failings. It has moved some researchers from scientists to advocates, where any exaggeration or cover up, as well as attacks on, exclusion of, or marginalization of opponents, can be justified on the basis of a perceived need to spur immediate government action to avoid a projected catastrophe. This indeed is an ends justify the means example right from the Socialist playbook.

Those disagreeing or even not completely on board with the agenda have been branded as Deniers, in an attempt to associate them with Holocaust deniers and discredit them. The debate over global warming, er... excuse me, climate change, has raged for decades with the SQR, politicians and the media shouting that the science is settled, there is a consensus, so shut up and sit down. Remember, consensus is about belief and is foreign to science, which should be based on unbiased facts.

Regardless of what you may have heard, the debate is far from over; there is no true consensus, which is meaningless to science anyway, and the science is not settled. The global warming hysteria has been promoted like a political campaign, with lots of hyperbole about the assumed severity and dire consequences if government action is not taken. It has become almost a religion to the faithful.

In November, 2009, emails[11] among leading researchers in the field were made public that seem to show a departure from an unbiased search for the truth to advocacy, complete with such things as suppressing research papers from those with different views, threatening to redefine peer review or boycott journals that publish them, massaging the data to fit the agenda and hide any embarrassing deviations, presenting proxy data as the real thing without explanation, mixing proxy and corrected or adjusted real data and destroying or hiding the raw data from other qualified researchers. Proxy data are things like tree rings, ice core gases, et cetera, that were used to estimate past temperatures.

Another tactic used by these advocates is to dismiss any conflicting information by, for example, claiming that the cited phenomenon was only local, not global. Thus, the Medieval Optimum (Medieval Warm Period) and the Little Ice Age are designated as insignificant or inconsequential local phenomena. But were they? The evidence indicates otherwise.

Since the first revelations, other errors have started to come to light such as IPCC [12] using unsubstantiated data from advocacy groups. Some researchers or officials have had to admit their role in the affair and others have either resigned or been reassigned by their organizations pending investigation into possible professional misconduct. In November, 2011, a new dump of selected emails[13] from the same time period has surfaced that verifies and deepens the misconduct and associated cover up.

Respected journals such as *Science*, the journal of the American Association for the Advancement of Science (AAAS), have become such advocates that they openly show their bias and vitriol in editorials, and probably also require the obligatory phraseology for publication of any article or report even vaguely touching on past or present climates.

Whatever the outcome of the debate, at least now there can be an open search for the truth. Computer models, of which there are several, are another problematic feature of climate change research and the conclusions reached. Each assumes that ALL of the factors affecting both weather and climate and the relative importance of each factor are known. The atmosphere is an extremely complex dynamic system, bound as it is to ocean, land, space, and sun. Although much good scientific information has been revealed through satellites and other tools, it is questionable whether we are anywhere close to completely defining the system.

CHAPTER 3

FALSIFIABILITY AND PARADIGMS

Science is the study of physical aspects of the universe that are reproducible, predictable, and that can be measured and observed. One-time phenomena, such as the origin of the universe or life, do not fall into the realm of science. Nor does absence of an alternative explanation, which does not call on a higher power, constitute verification of a concept. To be considered truly scientific, a theory or law must not only be verifiable by evidence and observation, but it must be falsifiable[1] and testable.

To be falsifiable means that there must exist a question or test that, if true, would render the theory false. Even if no counter-example is ever found, the question or test must exist. If no such question exists, then the theory is not truly scientific, it is a philosophy. Similarly, if it is impossible to test the theory by experimentation or observation, it cannot be verified to be probably true. Without being verifiable, falsifiable, and testable, the theory is a belief based on faith, and it cannot be considered fully scientific regardless of supporting evidence, ie, examples, models, or mathematics.

Evolution, origin of life theories, and cosmology are fields that easily fall into these categories. They have more in common with philosophy, religion or politics than science. Did life spontaneously emerge from non-living molecules and evolve into more and more complex life? Did the Big Bang really happen? Neither of these questions is verifiable, falsifiable or testable in any meaningful way. What question, discovery, or experiment could possibly make them falsifiable? We can give evidence that seems to support the theories, but only by interpretation of so-called proxy data —where one thing is measured in lieu of another, more difficult or impossible measurement and interpreted as meaning the other. Never mind that there are no known mechanisms for how they could have worked; these can be manufactured (surmised) as needed to fill in the clever story.

Survival of the fittest, the supposed mechanism of evolution, is not a mechanism at all, just a truism or tautology. By definition, those that survive are fit to survive. That doesn't answer the question of how evolution actually works. Describing WHAT is believed to have happened or assumed connections between fossils and living specimens does not tell HOW it actually happened or even THAT it did happen. What is the mechanism, the HOW? Not even DNA studies have answered *how* one species can be transformed into another, considering the different numbers and types of genes and the varying lengths and number of fragments of DNA. We know that even minor mutations are almost universally fatal, detrimental or result in sterility, so a beneficial random change must have been extremely rare, let alone a series of random changes that each gave a survival advantage and ultimately resulted in a new complex structure or process.

Even if the theories are true, there is no way to prove them false because we cannot go back in time. Any observation that seems contrary can be explained away as a special case or can be made to fit into the present theory by re-interpretation. For example, many animals that exist today have not changed much from those in the early geologic record, eg, scorpions, spiders, cockroaches, dragonflies, ants, crocodiles, turtles, frogs, sharks, gingko, ferns, cycad trees, jellyfish, some species of fish, crustaceans, bacteria, protozoans, and a long list of others, but that does not falsify the theory of evolution. Panchronic[2] species are merely interpreted as meaning that these creatures were exceptionally well adapted to their changing environments so that little or no change has been needed for their survival. Note that "needed" implies purpose, not random, accidental, mindless, molecular changes that fortuitously enhance survivability. It is interesting that evolutionists often describe such processes as learning or other-directed, purposeful, or mindful terms.

One example given to represent evolution in action is the development of antibiotic resistance in infectious bacteria. It is often described as microbes acquiring new genetic information as a means of survival; that somehow antibiotics actually caused the change. The language is that of purpose, whether or not it is assumed that the bacteria actually acquire new information, or whether it is assumed that already existing information is allowed to flourish when competing strains are eliminated by antibiotics. Note that the evolved bacterium is just a different variant of the same species, not a completely new type of bacteria, so it is a bad example of evolution, which is about changing into something else entirely (thus *Origin of Species*, not origin of varieties within species).

PARADIGMS, SIMPLICITY, AND AD HOC STATEMENTS

If science is true to its stated purpose of searching for the truth, when new facts are found that are contrary to a theory, the theory must be modified or replaced.

Science must also be willing to admit to its inability to prove an absolute or a negative. Theory should never be confused with fact, regardless of what the SQR may tell you. Other points of view and contrary evidence should be examined dispassionately for their validity. True science is never threatened by truth or contrary opinions, and open debate should be welcomed. Unfortunately, much of the practice of science has always done the opposite. This is because it is composed of humans that tend to protect their beliefs and intellectual territory in spite of contradictory evidence, especially if they feel threatened.

The business of science works within a paradigm – a worldview within which all theories must fit – and sometimes uses extreme measures to defend the established view. An observation that does not fit the theory must be made to fit by ad hoc[3] statements added to make it fit both the theory and the underlying paradigm. By adding many of these special case statements to a theory, the theory becomes more and more complex or convoluted, and, in the mind of the scientists involved, more and more true or accurate.

But is this assumption really true? Only in the sense that the new observation is made to fit the established theory and paradigm. A theory with numerous ad hoc additions may also indicate a dying theory that needs to be reformulated. Shifting a paradigm in science, as in other fields of study, is a monumental task that is usually met by great resistance from the advocates of the established dogma. These advocates ("experts") have a vested interest in maintaining the view that supports their life's work and publications. Scientists are only human and fall prey to all the usual pitfalls such as pride, emotional attachment, and defense of territory.

> Like the choice between competing political institutions, that between competing paradigms proves to be a choice between incompatible codes of community life. Because it has that character, the choice is not and cannot be determined merely by the evaluative procedures characteristic of normal science, for these depend in part upon a particular paradigm, and that paradigm is at issue. When paradigms enter, as they must, into a debate about paradigm choice, their role is necessarily circular. Each group uses its own paradigm to argue in that paradigm's defense…. [T]he status of the circular argument is only that of persuasion. It cannot be made logically or even probabilistically compelling for those who refuse to step into the circle.
>
> *- Thomas S. Kuhn, The Structure of Scientific Revolutions*

Occam's razor is a principle (read philosophy) of science that says, given more than one choice, the simplest solution is usually the correct one. On the contrary, adding more and more terms to fit new data into old theories is an anti-razor. Crabtree's Bludgeon is a satirical anti-razor which states: "No set of mutually inconsistent observations can exist for which some human intellect cannot conceive a coherent explanation, however complicated."[4]

This reminds me of an article, complete with sophisticated mathematics, published several years ago in the journal *Science*[5] that attributed the origin of crop circles to naturally occurring swirling down-drafts of air.[6] The journal only retracted the paper and apologized to its readers when two men admitted to starting the craze in England and demonstrated how they used boards and ropes. Given the elaborate designs that are now found, there is no way a natural atmospheric phenomenon can account for them. A human designer must be involved – or aliens according to another theory (read belief). Of course, true believers in a natural cause say only the simple circles are natural.

By the way, the biochemical processes of life within each cell appear to be examples of anti-razors. Instead of being a streamlined, straightforward process as was expected, each process is much more convoluted and complex than anyone ever imagined. This is true because each step of each process is tightly regulated by enzymes to insure that just the right amount of each product or energy source is present at each step. Metabolic pathways make Rube Goldberg contraptions look like models of simplified efficiency. This is beyond rocket science, the traditional analogy to illustrate complexity. Some of the simplest of these processes will be described in a later section. At least for biochemistry, the philosophy of Occam's razor, that simpler is truer, has not been verified.

A theory that must be made more and more complex is a dying theory that needs to be reviewed, modified or replaced and a new approach or paradigm shift may be needed. Next I would like to present two detailed examples of scientific paradigm shifts that illustrate some of the difficulties involved.

A classic example from history, as noted earlier, is the paradigm shift from the Earth-centered universe of Aristotle to Copernicus' sun-centered system. Because the Earth rotates on its axis, stars, sun and moon were observed to "move" across the sky. From observation, it was concluded that the entire universe revolved around the Earth. However, because the Earth and the planets really revolve around the sun in elliptical orbits, describing the motions of the planets to approximate the circular orbits that they had assumed, required orbits that involved loops upon loops (Ptolemaic epicycles) in which planets were seen to move in nearly circular loops across the night sky. This observation is really caused by the relative position of the Earth in its yearly orbit. Those holding the established paradigm saw nothing wrong with this picture. It was just one of those ad hoc hypotheses that made the observations fit the theory and the paradigm of the time. See diagram.

All astronomical knowledge before Galileo perfected the telescope had been through unaided observations. Through his astronomical observations, Nicolaus Copernicus, a Catholic cleric and scientist, formulated the theory that the Earth and the planets revolved around the sun.[7] His theory was widely disseminated across Europe and was accepted by most of the leaders of the Catholic Church, although some emphasized that it was only a theory, not a fact. Based on his astronomical expertise, he was even called upon by the pope to assist in correcting the clerical calendar. He demurred because he felt that more precise observations were needed, whereupon he made the needed measurements. Some years later those calculations formed the basis for the formulation of the Gregorian calendar that we use today. His six-volume work, *De Revolutionibus Orbium Coelastium*

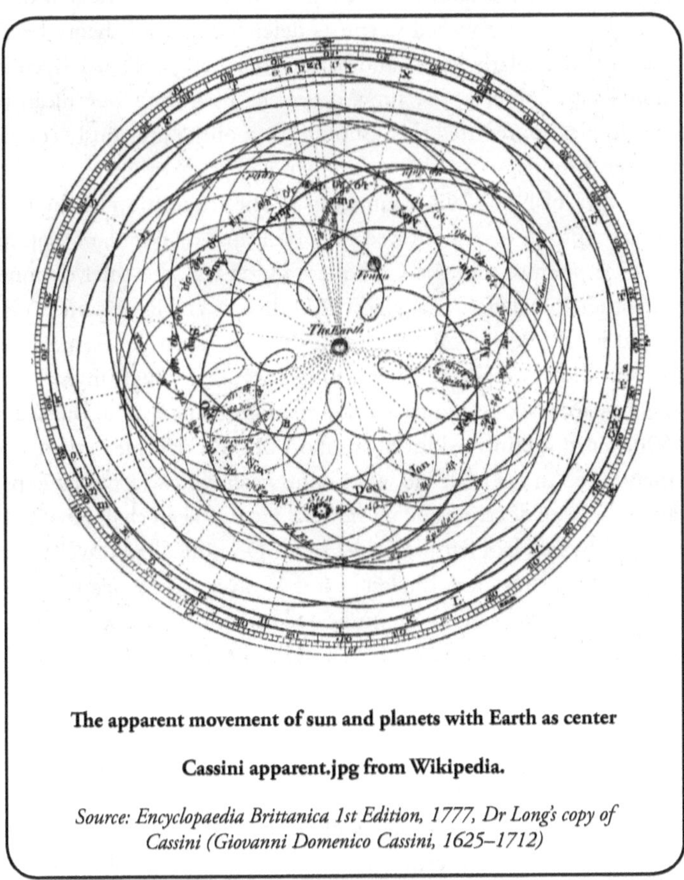

The apparent movement of sun and planets with Earth as center

Cassini apparent.jpg from Wikipedia.

Source: Encyclopaedia Brittanica 1st Edition, 1777, Dr Long's copy of Cassini (Giovanni Domenico Cassini, 1625–1712)

(Of the Revolutions of the Celestial Spheres), dedicated to Pope Paul III, was published in 1543, the year of his death. Both John Calvin and Martin Luther rejected the Copernican theory and some scientists ridiculed it. It had one failing. It did not quite hold up to reality because of the perfectly circular orbits that Copernicus used, rather than the more accurate elliptical orbits.

Although Copernicus' theory of a sun-centered planetary system had been accepted by many as a sensible alternative to the convoluted planetary orbits of the Earth-centered system, it fell to Galileo to give actual evidence that not everything revolved around the Earth, through observations of four moons revolving around Jupiter and by the phases of the planets Venus and Mercury. First published as *Sidereus Nuncius (The Starry Messenger*, 1610), Galileo's observations sparked fresh debate and study all across Europe in both scientific and theological circles.

Resistance came from many sources that vehemently defended the old paradigm, but many other scientists and clergy readily accepted the new direction – pun intended. Many established scientists defended the old theory because it was the basis of the scholarly books and papers they had published. Tyco Brahe, the most noted astronomer of the time, never accepted the Copernican theory but Johannes Kepler appreciated this simplification of the previously convoluted system and later formulated the theory involving elliptical orbits.

The true story of how Galileo ran afoul of the Inquisition of the Catholic Church is quite a bit more complicated than indicated by popular accounts that support the false premise that the Catholic Church was the chief opponent of scientific and other knowledge through the Middle Ages and Renaissance. Galileo's problems were more theological (and tactical) than scientific. Remember that Copernicus was widely respected throughout Europe although many disagreed with his theory. Contrary to modern accounts, Giordano Bruno, burned at the stake in 1600, was not punished for his support of the Copernican theory. It was not even mentioned in his trial. He was a contrarian who rejected large parts of Catholic and Christian doctrine and authority.[8] He had been charged with heresy as early as 1576 (twice), had left the Order of St. Dominic, had traveled through Germany, Switzerland, France, England, and Venice, railing against authority everywhere, and had also been excommunicated by the Calvinist Council and the Lutherans. Of course, that doesn't excuse the Inquisition for executing him for his beliefs. The point is that his scientific views had little or nothing to do with it.

Galileo's first two works, *Sidereus Nuncius (The Starry Messenger)* and *The Assayer* (1623) were supported by Pope Urban VIII, although with minor modifications requested by the Inquisition[9] that didn't sit well with Galileo. His first work described the discoveries of Jupiter's moons and Venus' phases as supporting Copernicus' sun-centered system. Galileo's troubles really started when, due to resentment of revisions requested by the Inquisition, he published the homily *Dialogue on the Two Chief Systems of the World* (1632), having the character Simplicio, a simpleton, as the defender of the Aristotelian Earth-centered theory and putting Pope Urban's requested revisions in his mouth. This insult to his long-time supporter was as if, tiring of Inquisition editorial requirements, Galileo was asking for a fight.

Presented to the Inquisition when he was under investigation for *Dialogue*, a letter Galileo had written years earlier to Benedetto Castelli, a Benedictine monk and former student, *The Authority of the Scripture* (1614), attempted to explain his views about how Copernican theory agreed with the scripture. In this letter, he surmised that scripture had been written for simple, uneducated minds using familiar language they could relate to, and that the real meaning of the scripture was often quite abstruse. Galileo was convicted and sentenced to prison for this religious interpretation, but his sentence was commuted to house arrest for life, where he continued his work and outreach in relative comfort.

The whole history of this case indicates that the so-called war against science by the Church has been overblown or misrepresented by the anti-God SQR. In fact, many early scientists were Catholic clergy studying at Catholic universities, which were the preservers and repositories of ancient and classical knowledge. The Catholic Church, whether you agree or disagree with many of its theological teachings, has always maintained universities, observatories, and other bodies supporting scientific research.

PARADIGM SHIFTS 2 & 3

A more recent case of fundamental paradigm shifts is represented by our changing view of the universe caused by a couple of important discoveries. Before the late nineteenth or early the twentieth century, the universe was thought to consist entirely of the Milky Way Galaxy. When telescopes were built that were powerful enough to discern individual stars within other galaxies, they soon discovered that Andromeda and other "nebulae" were really distant galaxies composed of stars. For the first time, it became obvious that ours was but one galaxy among many in a much larger universe. The fuzzy patches that had been assumed to be nebulae (meaning clouds) of glowing gas in or near our own galaxy turned out to be distant universes as large and complex as our own galaxy. This indeed was a serious paradigm shift that changed our picture of the universe forever. Being a fact directly observed, this paradigm shift met relatively little opposition. What is IS.

A second associated paradigm shift changed the assumed static universe into a dynamic, expanding one. The light from nebulae, now recognized as other distant galaxies, were known to have spectra that were shifted toward the red (longer wavelengths). The redshifted spectral absorption lines of nebulae had been observed as early as 1864 by Sir William Huggins, but, because they were thought to be nearby, the large shifts had been interpreted by astronomers as, among other things, unique elements or ionized states of matter not present on the Earth.[10]

Huggins was also the first to connect the spectral shifts of nearby stars to their speed using the Doppler effect. The Doppler effect is the compression or expansion of waves, eg, sound or light, by an approaching or receding source. A simple, though imperfect, example is the sound of a train whistle that is higher pitched (compressed) when approaching and lower pitched (expanded) after passing.

Edwin Hubble and others applied this to galaxies. Galaxies that appeared to be more distant than others also had their spectrum of light shifted more toward the red (stretched) end of the spectrum. The redshift /speed/distance relationship, ie, Hubble's Law, soon followed, amid considerable controversy that persisted for years about what the spectral shift really meant. Why would more distant galaxies be receding faster than nearer ones? Since light takes time to travel to us, the distance is also like looking into the past; the more distant, the older the light. At the time, it was natural to assume that older light might either be tired or that galaxies were actually traveling faster in the past and that the apparent expansion of the universe was slowing down.

Hubble and company thought it was due to expansion of the universe as predicted earlier by a solution of one of the models of Einstein's theory of general relativity, although it did not explain farther is faster. Fred Hoyle and others favored a theory known as the Steady State theory that stated that the universe is expanding but new matter is being continually created to fill in the empty spaces thereby maintaining a steady state.

Another theory by Fritz Zwicky, called "tired light" by its detractors, postulated that there is no expansion, but that light loses energy and thus shifts to longer (redder) wavelengths due to encounters with the gravitational fields of intervening matter. Each of these and other theories appeared to explain the redder is farther phenomenon, but Hubble's expanding universe theory eventually won out. There is still some controversy, and questions remain about what it all means. The SQR have accepted Hubble's expanding universe absolutely because it fits into their progressive view of the universe.

Hubble's Law says that the redshift of light from distant stars and galaxies, and therefore their assumed recessional speed, is proportional to the distance. The greater the redshift and thus their recessional speed, the greater the distance is assumed to be. The proportionality factor is the Hubble Constant. Since Hubble et al only dealt with nearby galaxies, a direct linear relationship attributed to Doppler effect was postulated. Certain types of stars, Cepheid variables, with the same variability period were assumed to be the same brightness everywhere and were used to support the distance to redshift relationship for nearby galaxies. Problems arose when more distant galaxies appeared to be receding faster than the speed of light, the theoretical limit of speed. Current theory is that space itself is expanding carrying galaxies with it like raisins on a rising loaf of bread. Since space is assumed to be expanding, the apparent (not real) speeds of objects would be cumulative, making more distant galaxies only appear to be traveling faster. When this program was run in reverse, it led to the Big Bang theory for the origin of the universe. Whatever the reason(s), redshift does seem to correlate to distance, though not in a simple linear way. We will discuss these theories in a later section.

The discovery of the relationship between redshift and distance and its interpretation as expanding space, caused a paradigm shift from a static, infinite universe to a picture of an expanding universe that had a beginning. When first formulated in 1929, Edwin Hubble came up with an estimate for the Hubble Constant of 500 km/sec/Mpc. (km/sec/Mpc means that the recessional speed in kilometers per second is proportional to the distance in Megaparsecs). See box below for more explanation of these terms.

The establishment of the value of the Hubble Constant is an example of defining and refining a theory to maintain a paradigm. Through many refinements over the years, the most recent estimate for the Hubble Constant is 77 km/sec/Mpc +/- 15%. Estimates have ranged from 40 to 90 in recent years. There is still some conflicting data and the debate is not over yet. Some measurements even suggest that the expansion is accelerating. A smaller Hubble Constant means objects are farther away for the same measured speed (read redshift). At 77 that is 6.5 times farther than for the original estimate of 500 for the Hubble Constant, so the difference in the Hubble Constant is important to a fundamental picture of the size of our universe.

The Hubble Constant is not falsifiable or testable to any significant degree except for the nearest stars in our own galaxy where distance calculations based on redshift can be checked by stellar parallax. Stellar parallax is the angular shift that an object moves against the background due to the size of the orbit of the earth. Although recent advances in satellite and optical technology have improved stellar parallax accuracy and pushed capabilities farther, we still can't get beyond our sector of our own galaxy. It says nothing about redshift to distance relationship for distant galaxies that are not bound to our galaxy by gravity. If other factors contribute to the observed redshift, then all bets are off. Such a discovery, if found, would require yet another paradigm shift in the future.

The expanding universe is an example of a scientifically based philosophical belief and is not fully science. Although it may be true, there is no reason to assume (believe) that speed relative to distance of faraway galaxies are related in any way to the motions of stars within our galaxy or even our local group of galaxies, which are all bound together by local gravity. Although there are scientific observations that seem to support it, because we cannot measure the actual distances directly, we can never be absolutely certain of their interpretation, nor of the absolute magnitude. This is a bit over-simplified, but the basic principle is sound. We'll look at this and similar subjects more closely in later chapters.

A FEW DETAILS:

Redshift is the shift to longer (redder) wavelengths in the spectrum of light from stars and galaxies that is thought to be caused by the expansion and/or the speed at which they are moving away from earth. To be more precise, it is measured by the shift in spectral gaps (dark lines) in the spectrum caused by absorption of light at given wavelengths by certain elements, particularly hydrogen. The greater the shift, the greater the recessional speed, and therefore the distance, is assumed to be.

Doppler effect is the compression or expansion of waves from an approaching or receding source. The classic example is that of a train whistle whose pitch is higher as it approaches and then is lower after it passes because the source of the sound is a little closer or farther away with each wave, effectively compressing or stretching them. When applied to light, it is bluer (compressed) as it approaches and redder (expanded) as it recedes.

Light year is the distance that light can travel in a vacuum in one year. The speed of light is 3×10^8 meters per second, so a light year is approximately 9.46×10^{12} kilometers or about 5.88×10^{12} miles, (9.46 trillion kilometers or 5.88 trillion miles).

Parsec is the distance to an object with a stellar parallax angle of 1 second of arc, which is 3.26 light years or 30.857×10^{12} or 30.857 trillion kilometers (19.178 trillion miles) distance. A Megaparsec is 1,000,000 times one parsec, or 30.857×10^{18} kilometers. 10^{18} is 1 followed by 18 zeros or 1,000,000,000,000,000,000 kilometers distance, so a megaparsec is 30.8 quintillion (or billion billion) kilometers!

Stellar parallax is the angular distance a star appears to move against the background due to the motion of the earth in its orbit. Knowing the size of Earth's orbit and the angle, the distance can be determined by triangulation. The smaller the angle, the more distant the object is. Even the nearest star has a parallax angle less than one second of arc. To give some perspective, a circle is divided into 360 degrees times 60 arc minutes per degree, times 60 arc seconds per arc minute or 1,296,000 arc seconds, so you can see that this is a very, very small angle. Although ever smaller parallax angle measurements are now possible with satellites and improved precision instruments, it is still useless for the nearest galaxies and even for more distant stars in our own galaxy. About 8 million stars are within approximately 200 parsecs (652 lt yrs.) currently measurable by stellar parallax.

MORE DETAILS:

At greater distances redshift is checked using so-called "standard candles."

Standard candles are assumed to be the same brightness everywhere, so smaller apparent brightness is assumed to be farther away, but there is no unequivocal means of verifying the results. Distances to more distant galaxies are even more uncertain because the standard candles are different. As a matter of fact, a series of seven distance yardsticks are used that may or may not correlate well with each other.

Standard candle is a term applied to a stellar object of (assumed) known brightness or other characteristics. In the case of nearby galaxies, the standard candles are Cepheid variable stars where absolute brightness is believed to be constant for those stars with the same variability frequency. (They blink at the same rate.) For more distant galaxies, a series of other standard candles are used. Each standard candle is useful within a particular distance window, usually without overlap. Counting from nearest to farthest they are: radar, stellar parallax, spectroscopic parallax (really fitting stars into a brightness and color vs. size relationship), Cepheid variables, Tully-Fisher galaxy brightness relationship, type 1a supernova, and redshift alone.

PATCHED TOGETHER FORMULAS

As an industrial scientist, I have worked with many talented engineers and scientists. When a manufacturing process has a problem, most engineers fix it by adjusting the system or adding a patch to minimize or eliminate the symptoms of the problem. They seldom look at the fundamental weaknesses of the entire process or design a whole new system from scratch. It would be too expensive in time and money to fix most problems if it required replacing the existing machinery or designing and optimizing a whole new system. Some patches create unforeseen new problems that also need patches. The result, over a period of time, can be a cobbled together process that is far from the original streamlined design, and that may be running outside the original specifications on the ragged edge of failure. The same can be said of any type of improvements when fundamental understanding of the process is incomplete.

Original designs and the scientific principles supporting them or the history of earlier patches and adjustments may be lost or poorly understood by those working on fixing problems. Eventually, no one fully understands why things were originally designed as they were or why they were changed, much less the consequences that yet another change may have. In addition to the cost considerations, sometimes engineers are so emotionally committed to the existing process and their own previous work that they are unwilling to consider alternative ways to fix problems. This is especially true if the answers involve doing away with cherished assumptions (beliefs). Although a better solution to the problem might be to remove all patches and set the process back to its originally designed conditions, this is seldom done, especially if critical fundamental understanding is ignored, lost or lacking.

This is also true of the business of theoretical science and research. It may be too expensive to pursue new avenues and propose solutions outside the established system of theories or paradigms. Careers and credibility have been destroyed when someone stepped outside the established norm. There is a great deal of pressure to conform, especially in academia, so patches are applied in the form of ad hoc statements. How many experimental observations have been ignored or discarded as either experimental error or crackpot schemes because they did not fit the accepted picture? I would venture to say that right at this moment, there are scientists sitting on ideas that they are afraid to put forth because those ideas do not support the status quo.

As an experienced industrial research scientist outside the halls of academia, I have no such limitations, so that makes me an ideal person to examine and question everything without limits or fear of reprisals. Even in industry, pointing out weaknesses in established beliefs can be difficult and troublesome. Most people take the attitude of "not invented here," which translates to "if it's not my idea, it's not good," and is one of the reasons for resistance to new ideas everywhere. Sometimes, it takes months before those in charge are ready to accept such suggestions, and sometimes that acceptance is accompanied by claiming the "discovery" as their own. I have encountered this several times in my 30+

year career in chemistry and material science. After the first couple of times this happened to me, I stopped objecting and let them take the credit so that the correct actions could take place to further the project. In my opinion, it doesn't matter who gets the credit as long as the right thing is done, but mine is far from the typical attitude. Credit and acceptance is very important to most researchers, who are often hungry for anything they can claim as their own discovery or that will result in a patent.

CHAPTER 4

CRITICAL THINKING AND LOGIC

How are we to know what is fact and what is belief? Are there some tools we can use to determine the veracity of a statement? Most certainly. First, we can ask "How do they know that?" (HDTKT). For example, if you read an account of what is going through the mind of a person as he dies or, say, an animal grazing in the wild, you have to ask the question "How do they know that?" Since we can't get inside the head of a person who has died or of an animal, how can we know their thoughts, emotions, and motivations? The account is obviously fiction or at least a guess based on a projection of what the author thinks they (or he) would be thinking.

In science, you will encounter similar situations. The media[1] is especially adept at presenting scientific theories as proven facts. At least many of the more scientific journals and books sometimes offer some weasel words[2] such as *probably, possibly, presumably, we think, it is believed, it is assumed,* et cetera. Note the weasel words I used in the last sentence: *many* and *sometimes,* indicating that even this is not always true.

I recently viewed a television program on the History Channel about homo erectus, a supposed evolutionary ancestor of man, that not only described what they ate and that they carried themselves erect to walk, but also presented a detailed picture of their society including a mother caring for her young and battles between rival groups. HDTKT ? They guessed.

It was presented in a way that would lead the uninitiated to believe that we really do know such details from the evidence. All of this was based on a few sketchy partial skeletons and a few fossil humanoid footprints. No tools, no remains of campfires, no cave dwellings, nothing. Not even a real link between the skeletal remains and the footprints other than the presumed age. While there may be some validity to such speculations, they are quite far from proven facts. Often, in science as in other fields, when evidence and facts are lacking or

> We shall see that adherents of the best-known theory have not responded to increasing adverse evidence by questioning the validity of their beliefs, in the best scientific tradition; rather, they have chosen to **hold it as a truth beyond question, thereby enshrining it as mythology**. In response, many alternative explanations have introduced even greater elements of mythology, until finally **science has been abandoned entirely in substance, though retained in name**.
>
> *- Robert Shapiro, Origins. (Emphasis added)*

inadequate, clever "just-so" stories are substituted that fit the status quo theories and paradigms. Each story is little more than a projection of the scientist's or author's own opinions and prejudices.

So, the first thing to do is to sort out known facts from the surrounding projections, speculations, pet theories and just plain wild guesses that are presented as if they are facts. It is a similar situation with reading or hearing a news account and determining what parts are facts and what parts are opinion, analysis and editorials, all of which are projections of the writer, not facts. To understand the difference, you may need to dig for some background information on why they think so. Is the assertion based on good, solid evidence or not? Does it defy any logical or physical laws? How do they know that? (HDTKT ?)

SOME UNIVERSAL RULES OF LOGIC

- Law of Non-Contradiction – A statement can't be both TRUE and FALSE. Something either IS or it ISN'T.

- Law of the Excluded Middle – A statement is either TRUE or FALSE; there is no middle or third choice. Something cannot halfway exist.

- Law of Causality – for every effect there is a cause. Everything that had a beginning or exists has an origin or cause. This is the basis of science. Science is the search for causes (truth).

- Principle of Uniformity – causes in the past operated in the same way they do in the present and will in the future. This is the basis of forensics, where events in the past are deduced from evidence pieced together. Based on this principle, astronomers assume that the laws of nature are the same everywhere in the universe.

INDUCTIVE REASONING AND ITS WEAKNESSES:

Science uses inductive reasoning to go from individual observations to general or universal statements. If we make the inductive statement, "All swans are white," we have made a decision that whiteness applies to all swans based on a large number of observations. However, if one non-white swan is observed, the statement is proven false. Even if you never see a non-white swan, you can still ask the question because you can never observe every single swan that may exist. In this sense, *it is falsifiable* because there could be a counter-example that would render the statement false – even if the counter-example is never found.

If we rephrase it as, "There are no non-white swans," there is no way to prove that a non-white swan does not exist somewhere in the world. Since it involves proving a negative, which is impossible, it is *not verifiable.* In that sense, the statement is a belief based on faith and can never be proven absolutely true. Universal (or "all") statements cannot be proven absolutely true for all cases. We can only really say, "All observed swans are white." This is a simple or naïve logical example, but the principle can be expanded to fit all scientific theories. Theories are sets of statements describing conclusions about observations. They are inductive conclusions based on specific cases to imply or induce general principles. Science can really only say, "Based on observations and experiments to date, theory X has been shown to describe the observations better than any other theory."

DEDUCTIVE REASONING AND ITS LIMITATIONS

Deductive reasoning starts with a general principle and seeks specific examples. Most deductive arguments are in the form: If independent statement A (premise) and dependent statement B are both true, then conclusion statement C is also true. Independent statement A is usually an inductive statement based on a body of knowledge. It draws its validity from numerous observations, which may or may not hold true in all cases as noted above. If A and B are true, then C must be true.

Example:

A. All swans are white (independent inductive statement),

B. John saw a swan (dependent statement)

C. The swan John saw was white (conclusion).

If either A or B are not true, then C may not be true either. Notice, I said "may not." In this case, C can be true if B is true even if A is not true. Swans may come in several colors, but John may still have seen a white swan on this occasion. Therefore, in this case, the truth of C is only dependent on John's testimony and is not a true conclusion based on deductive reasoning.

In deductive reasoning, C is necessarily true if both A and B are true and if the reasoning follows from A and B. Even if A and B are true, if B is not truly dependent on A, then you can't conclude that C is also true.

A better example would be:

A. All men are males

B. John is a man

C. John is male.

In this case, C is necessarily true if A and B are true. So, the rules that apply are: if you accept the premise A, and if the dependent statement B is truly dependent and is a subset of A, then C must be true. If the premise A is untrue or questionable or if B is not truly dependent and a subset of A, leaving some uncertainty, then C is not necessarily true. This is important to recognize when the logic does not follow in a seemingly logical argument.

THE LIMITS OF LOGIC

Logic is a philosophical means of ferreting out the truth, but must be used with caution. Deductive statements of general principles that are not well supported by or connected to the data are mere *assertions*, not arguments or proofs. When starting with a general principle from pure reason, because it is believed that it must be true, it is tempting to fill in the blanks exclusively with supporting data and ignore anything that is contrary.

Deductive arguments are only true when dependent statements are truly dependent subsets of the original premise so that they must be true. The original premise must also be true. In the following deductive argument, if premise A is true and dependent statements are true subsets of the premise, then conclusion C must be true. The caveat here is that first you must accept the validity of the premise a priori[3].

A. All men are male (by definition)

B. John is a man

C. John is a male.

Similarly, most inductive statements, going from specific observations and data to form general principles, cannot be proven absolutely and are only probably true, simply because everything is not or cannot be known. One exception is when all facts are known. The statement, "All of the letters on this page are black" is absolutely true because it is possible to know it completely from experience, ie, the data collection is already completely done. When used in the argument below, it is certain that both *A* and *C* are true because all facts are known.

A. All the letters on this page are black

B. John read this page

C. The words John read were black.

However, in the argument below it is not possible to know the color of every swan that has lived or will live in the world, so that *C* is only probably true because inductive statement *A* is only probably true from the knowledge (data) we have already collected.

A. All swans are white

B. John saw a swan

C. The swan John saw was white.

Science uses inductive statements to go from specific observations or data to a general principle. Most truly scientific theories are inductive and therefore cannot be known absolutely. Some scientific theories are deductive but may be based on assumptions that cannot be verified to be true. A theory may be so well researched that it has a high probability of being true. Scientific theories not well supported by data or not well connected to the data may be wrong while seeming true. Depending on your belief system or paradigm of how the universe works, you may consider it as a fact and not just a very high probability.

Remember, in arguments you must assume that the premises are true for the conclusion to be true. If you assume that the premises are absolutely true, not probably true, you are more likely to believe that the conclusion is also true. But watch this one. The premises may be true (or very probable) but the conclusion may not logically follow and therefore may be invalid. Example: (A) Everyone dies. (B) Everyone drinks water. (C) Drinking water causes death. Premises A and B are true, but the conclusion reached is definitely invalid. It assumes cause and effect, not mere correlation. An additional error in this case is that statement (B) is not really a dependent statement of (A). It is really a separate independent statement or premise. It looks like a valid deductive argument, but it obviously is not. Fake deductive statements like this are sometimes used by advocacy groups and government agencies for convincing the public of their cause. A recent example may be that of CO_2 and global warming. It goes like this:

A. CO_2, which is a greenhouse gas, is increasing in the atmosphere, some of which is manmade

B. The Earth is warming

C. Anthropic (manmade) CO_2 is causing warming with dire consequences.

We know that Carbon Dioxide (CO_2) is a greenhouse gas, but is there a direct cause and effect for present warming trends? What is the extent of the effect and how much is from other factors such as solar cycles, increased water vapor and clouds? How much of the CO_2 increase is manmade? Is increased CO_2 the cause of warming or is warming responsible for releasing more CO_2 from oceans (is CO_2 a leading or trailing indicator)? The matter is not so cut and dried as the "deductive" argument makes it appear. This is not to say there is not some validity to the argument, only that the argument is not a true deductive argument. A and B are independent statements. Greenhouse gases may be only one of the major or minor causes, the portion that is manmade may or may not be significant, the amount of warming may or may not be consequential. See Chapter 15.

Logic, like statistics, can be used (misused) to prove all sorts of illogical things. Zeno, the fourth-century BC Greek philosopher who founded the Stoic school, used logic to prove that all motion is impossible. It went something like this: To move across a room, we must first move half the distance; but before we can cross half the distance, we must first move half of that distance. This cycle is continued until the half-distance is so small that you cannot move at all. This is known as Zeno's paradox.

Similarly, you may have encountered in a mathematics or physics class at school the bouncing ball problem that goes something like this: A ball is dropped and bounces one foot high; each time the ball bounces, it only goes half as high. When does the ball come to rest? The answer is never, if you follow the logic of half of half of half indefinitely. Of course, we all know that a ball does finally come to rest, but, based on this logic, it infinitely approaches zero but never quite gets there. In mathematics, it's called an asymptote, where a value approaches zero but never fully attains it. In this case, the error is in the original premise because it does not take into account other forces such as gravity, friction, air resistance, thermal effects and molecular motion, that eventually overcome the bounce altogether.

We could also prove that life does not exist, based on calculations of the probabilities of each development step taking place in an unguided, random way. Calculations result in such low probabilities that they would be considered impossible. Life, indeed, is the "infinite improbability drive[4]." Life shouldn't exist, yet here we are. Obviously, we are missing some important information.

Does that mean we can't use logic to determine the truth? Not necessarily. I just wanted to point out how easily logic can be misused, so that you should be very careful in its use and interpretation. In arguments meant to convince people of a certain point, logical rules are sometimes bent to make the statements seem more true. This becomes a mere assertion, not a true argument. Some argumentation techniques are called logical fallacies. These are statements that appear on the surface to be valid arguments, but which are misleading due to some flaw in the logic. Let's look at a few of them that apply to science.

SOME LOGICAL FALLACIES

Appeal to Authority – (Argument from Authority) we touched on this earlier. If an authority, leader, or expert says it is true, it must be true. That conclusion all depends on the authority, his knowledge of the subject, his agenda, and his belief system. This argument goes: if Einstein or (fill in the blank) said it, then it must be true. Another form of this is the consensus of experts. Remember COWDUNG[5]? Yes, a whole cadre of experts can be wrong if the framework (or paradigm) in which they believe is false or incomplete.

Also, remember that being an expert in one field of science or being accomplished in any field does not qualify for expert status in another, sometimes related, field. We are all ignorant in more areas than those in which we are well versed. I wouldn't want a physicist or engineer performing surgery on me, even if he is world renowned. Likewise, I wouldn't want a surgeon to build a bridge. What makes a celebrity (or scientist or politician) more qualified than others to comment on issues of the day unless they are also experts in that field? Nothing.

Even experts can be wrong and their theories should never be sacrosanct against criticism or question. Could Einstein be wrong? Of course. Experts are not infallible, and treating them that way leads us back to protecting the status quo and COWDUNG. One example from my past is a professor of zoology who insisted that holes found in pastures surrounding the college were due to pocket gophers. He had spent years studying them in the southwestern United

States, so he could be considered an expert on pocket gophers. The college was in Virginia, well outside their range. The holes were due to groundhogs (wood chucks) which are common in the area. If you are a hammer, everything looks like a nail.

Appeal to Ignorance – (or argument from ignorance) This argument says that if a theory has not been proven wrong, it must be true. On the other hand, it can be argued that if the theory has not been proven true, it must be false. Or yet, if there is a lack of evidence for one theory, then another theory is assumed to be true. This is also called an **Argument by Lack of Imagination**. It can be applied to the arguments for macro-evolution[6]. In the sense that, because no one has come up with a different theory, which excludes God as a cause, then Darwin's theory must be true. While there could be some validity in it, this is a specious[7] argument. Absence of another explanation does not make the current one true. The theory must rest on the evidence, not on an absence of an alternative theory or evidence to the contrary.

Argument from Silence – (argumentum ex silentio) Similar to Appeal to Ignorance; absence of evidence to the contrary or a different theory is deemed to validate the argument.

Argument from Personal Incredulity – A theory is thought to be proven untrue because someone finds it personally offensive, unlikely or unbelievable, and consequently a preferred alternative is thought to be proven true. Also known as **Argument from Personal Belief** or **Argument from Personal Conviction.**

Appeal to the Majority – Consensus as proof. A commonly held belief must be true. Consensus is an alien concept to science because it is based on opinion, assumption or belief not fact. Those advocating manmade global warming use this one profusely. They say the argument is over; it is a proven fact based on the opinion of a supposed majority of scientists. Never mind that many of those included in the "consensus" know next to nothing about that field of study and many competent scientists in the field disagree or have doubts. Truth is never determined by committee vote or marketing or propaganda.

> In questions of science the authority of a thousand is not worth the humble reasoning of a single individual.
>
> *- Galileo*

Appeal to Emotion – Replacing rational arguments with emotional appeals. Propaganda is often laced with emotionally charged statements. Hitler was a master of this one.

False Cause or Non Sequitur[8] – Incorrectly assumes that one thing is the cause or explanation of another; an argument where the conclusion does not connect well with the premise. Big trucks do not beget little trucks, although big (adult) animals beget little (infant) animals. City buses do not necessarily cause a reduction in the number of cars, especially when many buses are nearly empty or are scheduled poorly.

Irrelevant Conclusion – An argument for one conclusion really proves a different one. For example, arguments against Intelligent Design as not being science can also be applied to Evolution itself.

Amphiboly – A sentence that can be interpreted in more than one way. Example: The statement, "He only said that" could mean "only he" or "only said" or "only that." Was he the only one saying it? Does it mean that he only said he would do it and that he didn't actually do it? Does it mean he said only that and not the other statements attributed to him? As you can see, what is being modified by "only" makes a lot of difference.

Equivocation – Using the same term to mean two or more different things; changing what is meant in mid-argument. This is commonly used in evolution's defense where micro and macro-evolution are used interchangeably to defend the theory. Well documented variability within species is used in a confusing way to imply the ability of species to change into other species.

Changing the subject – Arguing for one thing to prove another. This is another favorite tool in evolution's defense. Example: Using changes within a species as "proof" of evolution into an entirely different creature. See equivocation above.

Red Herring – inserting another unrelated factor intended to throw off the opposition rather than address the issue. Example: Citing the belief in seven literal days of Biblical creation to change the subject and/or discredit valid scientific objections to classical evolution.

Straw Man – arguing against a weaker proposition to imply winning a stronger one. Darwin set up his case for evolution by arguing against repeated special creations from nothing— God points and Poof! a new animal appears. The prevailing view was really that changes had obviously happened through the ages, but that there was no evidence for a particular mechanism, divine or otherwise. Darwin didn't provide a fact-based mechanism either, just opinions; but in light of this straw man argument, many found it plausible.

Ad Hominum Attack – attacking the opponent on personal grounds not related to the subject, including name calling or mischaracterizing the opponent's beliefs. Example: Grouping people together to imply guilt by association to invalidate their arguments. Again, evolutionists group everyone who argues against Darwinian evolution with young Earth or seven-day creationists. This is also called **guilt by association**. Another form of this is maligning the opponent's motives or character.

Begging the Question – The premise automatically assumes the truth of the conclusion in the statements to support it. Example: saying "Everyone knows—." "There is a consensus among scientists that—" or "It has been proven that—"

Complex Question - The statement contains the conclusion within it.

Wrong Direction – An argument in which the cause and effect are reversed.

Post Hoc – because one thing follows another it is assumed to cause it. Example: According to statistics, people who smoke, drink alcohol and engage in sex at younger ages die younger, therefore early smoking, drinking and sex cause premature death. It does not take into account that this may imply a philosophy of risk taking and that the stated activity may have nothing to do with the causes of premature death, such as by accident, suicide or diabetes associated with extreme obesity and sedentary lifestyle.

Insignificant or Complex Cause – one thing is assumed to cause another but it is only one, perhaps minor, part of a group of causes. Example: The assumption that man is the cause of global warming, when other factors such as changes in solar radiance, water vapor and clouds, methane from animals and decomposing vegetation, changes in surface reflectivity and the fact that we have been recovering from the Little Ice Age since the mid-1700s could be equally or more important.

Appeal to Motives: Prejudicial Language – value or moral goodness is attached to agreeing with the argument; conversely, assigning immoral or sinister motives to the opponent. Those opposed to the theory of manmade global warming are judged to have sinister motives that will harm mankind and the planet.

STATISTICS, GOOD AND BAD

Statistics can be a good or a bad thing. It all depends on how it is used. If I am testing a physical property of a production sample, and I know there is a certain level of expected variability in my product, I must test a greater number samples to be sure that the average and statistical variability actually represents the state of the process. Similarly, unless I also look at the individual data points for the range of variability, how the data is skewed[9], whether it has more than one mode[10], et cetera, I will not have confidence in the data being meaningful.

For instance, test results may consist of five hundreds and five zeroes. The average will be fifty, although none of the real values were near fifty. If I reported fifty as a true value of typical results produced by the process, it could lead to a failure to recognize a problem. The results are obviously bimodal and as widely variable as possible implying that either the process or the test method is not in control.

Now it may be that one of two shifts operating the process turns a certain switch off when they are running so that readings are not collected. It's like the proverbial hospital janitor unplugging the life support to plug in the vacuum cleaner. Of course, what could be worse is that the second shift reports the same data as the first shift without actually taking readings, thus hiding the problems. Care must be taken to properly collect, analyze and report statistical data, whether it is readings off a machine or surveys of political constituents. Both can be grossly wrong if the methods, the data and conclusions are not correctly associated. Beware of statistics unless you have all the facts and know the biases of those reporting them. I refer you to the classic book *How to Lie with Statistics* by Darrell Huff. On that same vein, two other books I found that might be useful are *How to Lie with Charts* by Gerald Everett Jones and *How to Lie with Maps* by Mark Monmonier[11].

EXPERIMENTAL DESIGN TO IDENTIFY IMPORTANT FACTORS

When examining a process to either fix a problem or improve the process, it is beneficial to do an experimental design. Most experimental designs use a matrix of all of the variables at several settings and combine them in various ways to isolate the important factors and to determine positive or negative synergy. Synergy is where two or more factors work together or against each other to affect the process to a greater degree than either would do independently. Remember the foam not film example? An experimental design was actually run to determine and test this formulation, but all factors were not considered. With experimental design, minor and major factors can be identified and interactions can be determined. Without it, the wrong factors or conclusions may be chosen or drawn.

Even a simple change must consider all factors that may affect it. But what happens if certain variables are not known or considered? Then erroneous conclusions may be reached. One example from my experience is as follows. Due to a change to new dyes that were sensitive to pH, (acid level), microencapsulation of dyes was changed from a urea-formaldehyde walls that require either high or low pH to form, to a melamine-formaldehyde process that reacts nearer neutral but not at high or low pH. Someone decided that we needed to eliminate any excess formaldehyde at the end of the process and decided to throw in some urea for that purpose. Since urea does not react with formaldehyde near neutral pH, the result was unreacted urea crystals deposited in the dryer and the exhaust stack, followed by clumps of urea crystals falling onto the film coating causing damage. In this case, the rigorous experimental design and testing were actually skipped, important factors were ignored and the change was incorporated directly into the commercial product. A very expensive mistake. The only benefit was free fertilizer in the form of urea vapors applied to the farm field next door.

So how is this related to scientific theories and their validity? As in experimental science, if all factors are not considered or the wrong emphasis or relationships are assumed, then the conclusions may be wildly wrong. Of course, in purely theoretical science it is not possible to experimentally vary levels of factors, so it is done through computer models. This is valid only for well-defined systems where all or most important factors are known. The problem with these models is that they *assume* they know all of the factors, their relative importance and how they interact. Models of global climate, continental drift, Earth's interior, star and galaxy formation, et cetera, assume a lot. Are they correct or even in the ball park? Maybe. Maybe not.

COMPUTER MODELS – USES AND CAUTIONS

As stated earlier, computer models are really sets of mathematical formulas representing the variables that interact to produce the effect being modeled. They have a lot in common with computer games and follow the rule GIGO (garbage in garbage out). Whether it is a model of the fluid flow patterns and energy profiles in a mixer with various types of blades and baffles or the weather or the climate, all computer models are basically the same. In making a model, scientists define the variables, determine how they affect the system and how they interact, define the weight of each variable in mathematical formulas and run the program to produce a set of results that can be represented by pictures or movies of the system in action.

If all variables are not known or are given the wrong emphasis in calculations or start off with the wrong values, or the formulas are off in any way, then the results will be wrong, sometimes grossly so. Many of the steps in setting up a computer model are based on assumptions about how the system works, what variables are important (and which can be ignored) and how the variables chosen are believed to affect it. If these assumptions and the beliefs they are based on are not accurate, then the result will be affected. Remember, there is a tendency to look for what we expect to find and ignore what we don't. A small error in input data or the way it is analyzed can be greatly magnified as the program is run over a period of time, the longer the time, the greater the error.

I know that most people are not equipped to analyze computer models to detect any built in errors, but we all can be cautious about believing the theory they illustrate without other corroborating evidence. Computer models are only as good as the initial assumptions, the data used, the weight they are given and the rules of the game employed to arrive at the predicted conclusions. Computer models are used in all kinds of areas to give legitimacy to the subject in question. Do not assume that a theory (belief) is more true just because someone has created a computer model. If it is based on the wrong assumptions (beliefs), if all factors are not known or some are ignored, misinterpreted or over- or under-emphasized, the model is little more than an illustration of the underlying belief system.

PROBABILITIES AS TOOLS

The probability of an event is commonly referred to as the odds. If the probabilities are sufficiently small, then the event is not probable. That does not mean it cannot occur, only that it has a very small chance of happening. The probabilities are seldom considered in quasi-scientific storytelling, much less in textbooks. The origins by random chance of the processes described are assumed to be probable. Remember this when reading any scientifically based narrative.

Let's look at a few simple examples to see how probabilities are calculated and the degree to which they could affect such philosophical "science" stories.

Example 1: The probability of turning up heads in a coin toss is 1 in 2 because there are only two possible outcomes, heads and tails. However, the probabilities of turning up 10 heads in a row are 1 in 2^{10}, that's ten twos multiplied together, or 1 in 1,024. You can see how multiple occurrences can quickly result in very small probabilities even for a simple choice of two. Now, if we had time to flip 10 coins 1024 or more times, turning up 10 heads in a row would become at least probable. It is a basic rule of probability that an event becomes probable if the number of tries is as great as or greater than the odds.

Example 2: The probability of drawing an ace of hearts randomly from a deck of 52 cards is 1 in 52 because there are 52 unique cards to choose from. The probability of drawing an ace of hearts from a deck of 52 cards each time on 10 tries, assuming the cards are all returned to the deck, is 1 in 52^{10} or 1 in $10^{17.160}$;

that's 144.5 x 10^{15} or odds of 1 in 144.5 quadrillion (or million billion). Most people would say that it is a practical impossibility because it is not likely that we could repeat the process 144.5 quadrillion times. If we drew 10 times and each time we drew an ace of hearts, we might say it is a miracle or that the fix is in and it is a magic trick.

Example 3A: Now suppose that we want to know how many combinations there are for three letters or numbers without repeats. Using factorials, the answer is 3! or 3 x 2 x 1 = 6 because the number available is reduced by 1 each time. For the letters A, B and C, this is ABC, ACB, BAC, BCA, CAB, and CBA.

Note that 3! means 3 factorial. A factorial multiplies all digits from 1 to the target number together so that 3! = 3 x 2 x 1 = 6

Example 3B: If repeats are allowed in Example 3A, it becomes 3^3 = 27, because each letter can be used any number of times. The result is AAA, AAB, AAC, ABA, ACA, BAA, CAA, BBB, BBA, BBC, BAB, BCB, ABB, CBB, CCC, CCA, CCB, CAC, CBC, ACC, BCC, ABC, ACB, BAC, BCA, CAB, CBA. This is similar to the card draw example because we are using all of the letters on each try.

Real World Example 1: There are 26 letters in the alphabet. How many seven letter combinations are possible? (repeats allowed)

Answer: 26^7 = 8.03 x 10^9 (or 8.03 billion)

What is the probability of the word *America* forming at random by chance?

The answer: 1 in 8.03 billion. It is just conceivable that a fast computer, given enough time, could spit out a long enough sequence of combinations to hit on the word *America* by chance, so it becomes probable given enough tries. At 100 combinations per second, it would take 2.5 years to make all possible combinations.

Real-World Example 2: Chains of amino acids selected from 20 different left-handed amino acids[12] make up the proteins in all living creatures. Enzymes are special proteins that facilitate biochemical processes inside cells. If a typical enzyme is composed of about 200 amino acids, how many combinations are possible?

Answer: 20^{200} or 10^{260}. That's a 1 with 260 zeros after it. Note that this is an over-simplification or perfect example. It assumes not only the perfect conditions for each reaction, but also a readily available supply of all of the left handed amino acids, and the absence or sorting out of all of the right-handed amino acids and any other interfering molecules.

What is the probability of one specific sequence forming at random by chance?

Answer: 1 in 20^{200}. If we could sequence amino acids that many times, we might come up with the one enzyme we want, but that's just one enzyme. If the universe is 13.7 billion years old, there have been 4.32×10^{17} seconds since it began. We would need to make 231.4×10^{180} attempts each second since the beginning of the universe to make it plausible—that is, so that the number of attempts is in the same ball park as the odds. All of this assumes that all of the left handed amino acids are available in one place and the conditions are right for the reaction to occur. But that is just for one specific enzyme. There are thousands of enzymes in living cells so we would have to raise that by the number of enzymes needed. For instance, that would be $(20^{200})^{3000}$ for a simple cell containing 3,000 enzymes. It appears to be a practical impossibility. We will talk about these and similar odds in a later chapters on the origin of life and similar subjects.

The bottom line is that, given very, very small probabilities, if it is possible to perform enough tries to be similar in magnitude to the probability, (as large or larger than the odds), we might just hit on the desired combination. Or not! But remember this: even given these infinitesimal odds, each try has an equal chance of producing the desired result by sheer luck! You might do that once, but could you do it thousands of times by sheer luck? As in the example of drawing 10 aces of hearts in a row, it is unbelievable to assume that it could happen by chance, aka luck. That makes it effectively impossible and differs little from a miracle. Yet, we are here!

William Dembski, in *The Design Inference*,[13] made the case for discounting claims of random chance or regularities, caused by known laws as explanations if the outcome is both specified complexity and has sufficiently small probability, given all possible trials. Specified complexity is one of the arguments for intelligent design in nature. When examining any scientific theory, it is wise to ask how probable the proposed sequence of events is. If the probabilities of its occurrence are ridiculously small, then there is reason to doubt or at least question the scenario. Maybe there is another, more probable solution or unknown principle that has yet to be defined.

CHAPTER 5

MAGIC WORDS, MAGIC, MIRACLES, AND MAGICAL THINKING

MAGIC WORDS AND PHRASES

Magic words and phrases are those that are thrown in to make you believe something has been explained when it hasn't; it has only been labeled. A magic word can be a label given to things or processes that are currently unexplained by science. A magic word or phrase side-steps the issue and masks ignorance while seeming to explain something. Definitions and descriptions also can be magic when they are used differently to suit the current argument. A description of a process or thing does not necessarily explain it either. Be cautious when such terms are used. Ask yourself if it really explains anything or just labels it in a way that seems to explain it. What is *thought, gravity, instinct*? When we hear such terms, we feel like we understand something, but what do we really know? We really understand them in only the most superficial, descriptive way.

SHORT LIST OF MAGIC WORDS:

Processes:	**Things:**
Evolution	Species
Instinct	Big Bang
Relativity	Culture
Metamorphosis	Society
Gravity	Diversity

Vacuum Fluctuation	Space-time
Emergence	Black hole
Consciousness	Multi-verse
Mind	Vacuum Energy
Life	Gene
Uniformity	DNA
Tolerance	
Warp	

MAGIC, TRICKS, AND MIRACLES

Magic is really hidden technology or knowledge. The observer is led to believe he is seeing one thing, when something entirely different is being done. In a classic magic act, sleight of hand and other tricks are used to give the illusion of seemingly impossible feats. When magic tricks are explained, the illusion collapses with understanding.

If you believe that miracles do happen, miracles can be considered to be like magic although their purpose is not specifically designed to trick the observer. In a universe governed by physical laws, miracles imply the existence of some unknown factors that are involved but that are not evident or understood by the observer or recipient. A miracle appears to break the laws of the universe, but may really be attributable to hidden knowledge or unknown factors introduced to accomplish that which is seemingly impossible. Because it is based on hidden technology or knowledge, it does not require breaking any laws to appear to do so. A savage may think a recorded voice or image is magic or a miracle because he does not understand the technology. Similarly, miracles may be beyond our understanding while still obeying all of the laws of nature.

Would a creator invent laws to govern the universe and then break or bend them to accomplish a specific purpose? Maybe. Who are we to say what such a being would or would not (could or could not) do? However, in my opinion it would only be necessary to use those laws in unique ways that are beyond our understanding. Those who argue against a Creator-God who is involved with us personally tend to define miracles as breaking the laws of nature. As you can see, this is not necessarily the case. Their definition is faulty because it is based more on bias than logic. By the same token, because miracles are not reproducible or predictable, those same critics tend to dismiss as exaggerations, fabrications or myths, historical reports of miracles, even those witnessed by a great number of people and that are well-documented.

Do miracles happen? That is mostly a matter of belief or philosophy. It is not a scientific matter in the sense that it is not predictable or reproducible and we cannot test, verify, or falsify it. We are left only with a description of an isolated incident as told by those who witnessed it but with no known logical explanation. As with any eyewitness report, different people witnessing the same thing may remember it differently. Such evidence is anecdotal and is considered less reliable than hard evidence. There are enough documented cases of miracle cures that are unexplainable by medical science, such as the spontaneous disappearance of cancer, to believe that miracles really do happen. However, due to their capricious nature and lack of reproducibility, they cannot be fully understood by science.

Obviously something happened; some new factor or process occurred that we don't understand. If you accept that such things really do happen, they are still outside of science for the reasons stated above. Are they predictable? Can we make it happen again or determine after the fact what happened and why? For so-called miracle cures, the answer to these questions is no. If the answer is yes, then they are not considered miracles but are considered an application of medical science.

MAGICAL THINKING

What I call Magical Thinking is often practiced by those who push an agenda, whether it is scientific, political or cultural. What is Magical Thinking? It is the belief in an outcome that has never been demonstrated to be true in the past. For instance, those who believe in the theory of Darwinian evolution will go on and on about the overwhelming evidence for evolution, although it has never been observed in nature or been demonstrated by a direct link between any living and fossil creatures over the last 150 years of study. A Socialist will declare endlessly that free enterprise and our representative republic are exploitive and that a utopian state is possible without becoming both poorer and a dictatorial police state, although each time it has been attempted over the last 400-odd years it has failed to deliver as promised. Then they will declare that this time will be different; we are the ones who know how to do it right. This is nothing more than Magical Thinking.

A cosmologist will prattle on about multiple universes, ten dimensions and the Big Bang, without any demonstrable proof other than mathematical calculations and simulations based on a nested set of assumptions. It is as though loudly repeating a belief enough times will make it so. I am not necessarily saying all of these beliefs are entirely wrong, just that their basis is questionable at best. Such beliefs should be couched in lots of conditional words such as *possibly*, *probably*, or *it is believed*. Otherwise, declaring beliefs that are uncertain in boldly certain terms is nothing more than Magical Thinking.

PART 2

OVERVIEW OF "LIFE, THE UNIVERSE, AND EVERYTHING"[1]

CHAPTER 6

TOUR OF THE UNIVERSE, SOLAR SYSTEM, AND EARTH

We live in a universe of extremes, from vast distances, huge galaxy clusters and extreme energies to extremely tiny atomic and subatomic particles and infinitesimal energies. Living things, too, range from the huge to the microscopic, from the ages old to the ephemeral, from the powerful to the powerless in almost infinite forms and extreme complexity. Let's take a tour of the wonders of our reality. Since the subject is so broad and complex, I will attempt to give you an appreciation of its wonders without bogging you down in any more details than necessary. When more details are needed I may put them in endnotes, boxed comments, or appendices. Since some of the facts are still debated, I will stick to the status quo explanations for our tour. If you find the details in some of the sections of this chapter too tedious, you can just skip over them to the next section or skip the chapter altogether. However, if you can wade through them, you will gain an appreciation of each level of complexity that can help you in later chapters.

TOUR OF THE UNIVERSE

The universe is composed of vast stretches of empty space sparsely dotted with galaxies. Each galaxy is composed of billions to a trillion stars, some of which have planets, as well as copious amounts of interstellar dust and gases. Galaxies assume several forms such as elliptical, irregular and spiral discs with central masses and/or barred centers. Andromeda, our closest neighbor, is a spiral galaxy composed of a trillion stars. Our own galaxy is a barred spiral galaxy, perhaps as large as Andromeda.

There are hundreds of billions of galaxies arrayed in clusters such as our Local Group, the Virgo Super Cluster, or the Great Wall, a super cluster that is the largest known structure in the universe. Clusters surround huge, dark voids where no galaxies are apparent. The standard model[1] is of an ever-expanding universe that started with a gigantic "explosion" known as the Big Bang, which created all of the matter, energy, and space in the universe and even time itself about 13.7 billion years ago. The universe has been expanding ever since.

How big is it? Suffice it to say it is really, really, mind-bogglingly BIG! So big that no one can really wrap their mind around it. No one really knows how far it extends or whether there is an end to it, but currently the farthest object detected[2] is a small galaxy thought to be at least 13.2 billion light years distant, just 500 million years after the beginning of the universe, according to the Big Bang theory. A light year is the distance light travels in a year in a vacuum, so the number of light years is the distance and also the time it took for the light to travel to us. At 9.46 million million[3] km (5.88 million million miles) per light year, the observed distance to the farthest object would be 125 million million billion[4] km (78 million million billion miles).

When the farthest galaxy was detected, we saw it as it was 13.2 billion years ago, and if the responsible object is receding at a constant rate, it would now be farther away by how far it could travel in 13.2 billion years. The apparent speed is an ever-increasing rate the farther back in time (and distance) you go. This is a bit of an enigma since the universe is supposed to really be expanding at a constant rate[5]. This acceleration into the distant past is attributed to the very space between the galaxies expanding and thus exaggerating the apparent, but not real, speed of recession the farther out we go. Since the observable universe is assumed to be uniform in all directions, the diameter would be greater than 26.4 billion light years (radius of 13.2 times 2). With each advance in telescope technology, the observable universe has "grown" by significant amounts. Beyond that, there is the unobservable universe.

To give some perspective on size and distance, while our Milky Way galaxy, with perhaps a trillion stars, is a mere 100,000 light years in diameter, it is 2.5 million light years or 25 times the diameter of the Milky Way to the nearest spiral galaxy, Andromeda. By comparison, the average distance from Earth to the sun is approximately 93 million miles or 8.3 light minutes. How about this for comparison? The distance from New York to Los Angeles is about 3,000 miles (4,830 km), or 0.016 light seconds (16 light milliseconds). Sorry for the tedious details, but it is the only way to illustrate the size of the observable universe, which may be a mere fraction of the total size.

Let's see if we can put some perspective on these distances by an analogy. If we represent the Milky Way galaxy by 1 cm (0.4 inch), the distance to Andromeda would be 25 cm (10 inches). The diameter of the local group of galaxies would be 1 meter (100 cm or 39.4 inches). The diameter of the Virgo Super Cluster of galaxies would be 11 meters (1,100 cm or 43 feet) and the distance to the farthest object detected would be 1.32 kilometers (0.81 miles), ie, 13.2 million cm. The entire solar system, including the supposed very distant Oort cloud of comets, is less than 1/100,000[th] of the diameter of the Milky Way galaxy. So in this example, the diameter of the solar system would be represented by about 10 microns, (micrometers), or about twice the size of an average large bacterium, (assuming that the galaxy is represented by 1 cm).

If we try to carry this down to the relative size of the earth or of humans, the analogy breaks down into confusingly tiny numbers, so let's use a different analogy. If we let the solar system, with a diameter of 7.5 trillion km, be represented by 1,000 km, then the diameter of the earth would be 1.7 microns, about the size of a small bacterium. (That's 0.0017[th] or 1.7 ten thousandths of the solar system diameter.) Again, the analogy breaks down when we compare the height of humans to the diameter of the Earth, since the average height (170 cm) is 1.3 ten millionths the diameter of the Earth (12,750 km).

Since the sun comprises over 99 percent of the mass of the solar system and Jupiter and Saturn account for 90 percent of the remainder, an alien spaceship approaching the solar system might describe it as a star with four gas giant planets and debris, with the Earth as part of the debris[6]. Similarly, an alien looking at the Earth from space would consider the entire biomass, including man, as a relatively minor feature compared to the size of the Earth or its oceans or continents. I hope this helps to put in perspective the size of the universe and the relative size of our galaxy, our solar system, our planet, and us in it. Notice that I did not say the significance. Size and significance are two different things.

In order to make it seem ridiculous for man to think himself worthy of the attention of any Creator-God the Earth is often described as an unremarkable planet orbiting a rather ordinary star that is located far out on one of the spiral arms of an ordinary galaxy that is but one in billions. This is a specious argument because it is based on our presuming to know how a creator-god would logically think and behave. HDTKT[7]? Beats me! However, it is conceivable that any being that can create such ordered complexity could be capable of anything, including contact with us.

Does the size and complexity of the universe mean we are but one example of billions of similar earths with abundant, perhaps intelligent, life and are therefore not unique? Maybe, maybe not. Does this mean the Earth and life as we know it are insignificant or inconsequential or are but a happy accident? The answer to that depends on whether you are a materialist atheist or not. This argument for man's insignificance is often used by those who would claim the universe has produced life completely at random without the need for a Creator or Designer, but it is a specious argument, as stated above. It is merely an opinion, and it certainly is not science.

The rarity or abundance of life in the universe can be interpreted either as an affirmation of atheism or an affirmation of theism. It proves nothing; but only reveals the belief system of the originator. For the atheist and the theist alike it is a theological argument. Rarity can mean special, or not; abundance of life elsewhere can mean we are ordinary and the result of chance, or not. It all depends on your philosophical perspective. Science says nothing either way.

We exist. What is IS, whether it was designed by a Creator-God or came from a hairball coughed up by a giant cat, or was sneezed from the snout of the great green Arcelsiezier[8] or just appeared from nothing in a giant explosion unaided by anything. All of these scenarios are equally unfounded, unbelievable, fantastic, and weird. They are philosophy or pure speculation, and certainly are not science. But I digress. Back to our tour.

How much matter is there in the universe? The sizes of the galaxies pale in comparison to the vast distances between them. At least astronomers think the space between galaxies is emptier than the best vacuum we can produce on Earth. The universe is mostly empty space, so that averaging all the matter estimated to be in the galaxies and intergalactic space, it results in an average density of 1 hydrogen atom in 4 cubic meters of space, roughly a 1.6 meter or 5 foot cube. Doing the math based on the diameter of the hydrogen atom (2.2×10^{-10} or 0.00000000022 meters), there is roughly 7×10^{29} or 700,000 trillion, trillion times as much space as matter[9]. That's practically and effectively empty! Only gravity, which organizes the matter into galaxies, clusters, stars, and other bodies makes any of it more than just sparsely scattered gas and dust in the vast vacuum of space.

One figure given for the estimated total number of atoms in the observable universe is 1×10^{90}. That's 1 with 90 zeros after it, a truly inconceivable number. At 1 atom per 4 cubic meters, that would be 4×10^{90} cubic meters in the observable universe. The detectable matter in the universe, by number of detectable particles, is approximately 90 percent hydrogen, 9 percent helium and the balance is all the other elements combined.

It is thought that as much as 90 percent of the matter and energy in the universe are undetectable dark matter and dark energy, which are totally undetectable so far. The fact that the speed of rotation of stars around the center of galaxies does not decline at greater distances from the center as much as expected by theoretical calculations seems to indicate a massive dark matter halo; and there is evidence of an accelerating expansion of the universe that may be caused by dark energy. Dark matter is thought to slow the expansion of the universe through gravity. Dark energy is thought to promote expansion.

What is dark matter? No one knows, but two types are hypothesized: MACHOs or Massive Compact Halo Objects, probably composed of ordinary matter such as small black holes, failed stars called "brown dwarfs" or solid chunks of heavy elements, and WIMPs, Weakly Interacting Massive Particles of unknown composition, though none have been detected – and thus are "dark." Since WIMPs are thought to interact only through gravity, they could not be ordinary matter and are called "exotic."

Unobserved/undetectable matter need not be exotic, just cold, solid, ordinary matter that we can't see. The clouds of dust and gas in our galaxy, of which there is quite a lot, is only detectable because stars within them make them glow or by reflection of the light from nearby stars or by obviously blocking our view of stars and glowing gas and dust behind them. Outside of our galaxy in intergalactic space, these means of detection are almost nonexistent, so most of the vast amounts of dust and other solid objects that could be there would be undetectable except for blocking or dimming the light from distant galaxies. Hmm, clusters of galaxies around vast dark voids... it makes one wonder.

TOUR OF THE SOLAR SYSTEM

The Sun is 93 million miles from Earth. It is 865,121 miles (1,391,980 km) in diameter and weighs 2.2×10^{27} tons (2×10^{30} kg), which is 330,000 times the mass of the Earth. Its volume is over one million times that of Earth. It has a magnetic field, an atmosphere and produces a solar wind of radiation and particles blown out into space. It is composed of hot plasma (ionized gas), mostly hydrogen. Like all stars, it is fueled by fusion of hydrogen atoms in its core. The sun contains 99 percent of the mass of the entire solar system. The solar system has eight planets[10], several minor planets, numerous moons, asteroids, comets, and dust. Proxima Centauri is the nearest star at 4.2 light years away. The solar system orbits the Milky Way galaxy about halfway between the outer rim and the center in about 240 million years.

The Earth is the third planet from the sun and is the only one with a temperature range that allows it to have appreciable amounts liquid water supporting the only known life. It is 7,690 miles (12,750 km) in diameter and weighs 6.58×10^{21} tons (5.97×10^{24} kg). It has a magnetic field that protects us from harmful gamma and X-ray radiation as well as particles from the solar wind; it also has an atmosphere, extending up to 560 miles (900 km), whose constant generation of high altitude ozone from oxygen by solar radiation protects us from deadly levels of ultraviolet (UV) radiation. The Earth is tilted by about 23° from the plane of its orbit (ecliptic), which creates our seasons and widens the area of its surface where animal and plant life can thrive. See Tour of the Earth below for more details.

The other planets are hostile to life as we know it due to factors such as extreme temperatures, low gravity, absence or scarcity of liquid water, and lack of a breathable atmosphere. It is possible that some extremophile[11] bacteria could live on other planets. The moon and Mars are deemed the best candidates for exploration and colonization. For man to live on the moon or Mars, we would be required to take everything with us, including air and water, as well as shielding from harmful radiation, low pressure, and extreme temperatures. The low gravity (less than 40 percent of Earth's) would quickly deteriorate our bones and muscles and cause other harmful physiological effects, while its lack of significant atmosphere and magnetic field would expose explorers to dangerous levels of radiation. Returning to Earth after living for extended periods in low gravity could be debilitating or fatal.

Mars has no magnetic field to deflect harmful radiation and particles from the solar wind, which may have stripped away most of its atmosphere. The atmosphere of Mars, composed of 95 percent carbon dioxide, is so thin and so low in oxygen that it does not provide protection from harmful UV rays. Atmospheric pressure is 0.6 percent of Earth's atmospheric pressure, so that pressurized suits and habitats would be needed to prevent our bodies from exploding and our blood from boiling in the low pressure. Mars is hardly the desirable destination for colonization being advertised today, unless people are prepared for a one-way trip and a lifetime commitment to living inside pressurized containers that recycle everything.

SHORT DESCRIPTION OF SOLAR SYSTEM BODIES

- Mercury, nearest to the Sun, is very moon-like in size and description, is covered with impact craters and has practically no atmosphere.

- Venus is nearly the size of the Earth, but has a surface hot enough to melt lead, a thick, toxic atmosphere of carbon dioxide with sulfur dioxide clouds, little or no protective magnetic field, and is covered with impact craters.

- Earth has a thick atmosphere of roughly 78 percent nitrogen, 21 percent oxygen, traces of argon, carbon dioxide and other molecules, and it carries about 1 percent water vapor. The Earth's thick atmosphere and greenhouse gases[12] raise and regulate temperatures within a narrow range[13] that allows for abundant liquid water needed for life. It has a magnetic field that protects us from much of the harmful radiation from space and from the solar wind stripping away its atmosphere.

- Earth's Moon has no atmosphere and is covered with impact craters. Its orbit and rotation are the same so that one side always faces earth. The moon has practically no magnetic field and only one sixth (16.7 percent) of Earth's gravity.

- Mars has an atmosphere that has only 0.63 percent of Earth's atmospheric pressure, mostly carbon dioxide (approx. 96 percent), and is covered with impact craters, huge volcanoes, and rifts. It has no protective magnetic field and is very cold and dry with temperature swings from -143°C (-225°F) at the poles in winter to 35°C (95°F) at the equator in summer. Its gravity is only 37.6 percent that of Earth.

- Jupiter is the largest of the four gas giant planets and is composed mostly of hydrogen and helium. Its cloud tops are in constant turmoil, with gigantic storms tinted reddish by ammonia, etc. It has four large moons, Ganymede (largest in the solar system at 3,722 miles or 5,265 km diameter), Calisto, Io, and Europa, as well as 62 small moons, and a thin ring of ice and dust.

- Saturn is the second-largest gas giant, composed mostly of hydrogen, with one large moon, Titan, second-largest in the solar system at 3,200 miles or 5,150 km diameter, with atmosphere and possibly methane lakes; four medium sized moons, Rhea, Iapetus, Dione, and Tethys, 57 small moons, and a complex series of rings composed of dust and ice.

- Uranus is a gas giant composed mostly of hydrogen but tinged greenish blue by methane, which rotates at a tilt of 98° to the ecliptic (plane of the solar system) which means it appears to revolve around the sun on its side with first the north and then the south pole facing the sun. It has five large moons, Oberon, Titania, Umbriel, Ariel and Miranda, 24 smaller moons, and a thin ring, most of which orbit near its tilted equator.

- Neptune is a gas giant composed mostly of methane and ammonia, which tint it blue. It has three large moons, Triton, Proteus, Nereid, 10 smaller moons, and thin rings. Because Triton has an orbit that is very eccentric and tilted, and orbits in the wrong direction (retrograde), it might have been a captured asteroid or comet.

- Pluto is an icy minor planet with one large moon, Charon, and 5 smaller moons. It has the most eccentric orbit of all planets so that some of the time it is closer to the sun than Neptune.

- Asteroids and minor planets are found throughout the solar system, but most of them orbit in a region between Mars and Jupiter called the Asteroid Belt, in a wide range of distances from the sun, from about 2 to 3.5 AU (astronomical units or Earth distances). Their orbits range from nearly circular to very elliptical and from near the ecliptic (plane of Earth's orbit) up to 30° tilt. They range in size from Ceres at 584 miles (940 km) to dust grains. There are 26 larger than 60 miles (100 km) diameter and there may be as many as a million over 0.6 miles (1 km). The total mass is thought to be about 4 percent of the moon's mass. Other asteroids orbit near or cross Earth's orbit. Every day the Earth is bombarded with 50 to 100 tons of meteors, ranging from dust to baseball size, most of which burn up in the atmosphere.

- There are also groups of small asteroids that share Jupiter's orbit at 60° before and after the planet, called Trojan groups, which are held there by a balance between the sun's and the planet's gravity. Other planets, including Mars, Neptune, and Earth, have similar though smaller groups. There is another belt of asteroids and minor planets beyond the planetary orbits called the Kuiper Belt, ranging from about Neptune to 50 AU distance. It is the inner part of the Scattered Disk of objects which extends to, perhaps, 100 AU, with orbital tilts up to 40°, and is thought to be the origin of most short period comets. Beyond that, comets are thought to reside in the unconfirmed Oort cloud that may be spherical since these long period comets can approach the sun from any direction.

TABLE OF SOLAR SYSTEM PHYSICAL CHARACTERISTICS

Solar System Body	Distance to Sun in earth units (AU) earth = 1	Diameter in miles (km)	Temperature, F (C)	Gravity in earth units (earth= 1)	Day/ Year (as earth hrs, days, yrs)
Mercury	0.4	3032 (4879)	-300 to 800 (-180 to 430)	0.4	58.7d / 88d
Venus	0.7	7,521 (12104)	860 (460)	0.9	243d / 225d
Earth 1 moon	1	7,690 (12,750)	-128 to 134 (-89 to 57)	1	23.9hr / 365.25d
Moon	0.0026 (from Earth)	2,159 (3,475)	-247 to 221 (-155 to 105)	0.16	27.3d / 27.3d
Mars 2 moons	1.5	4,220 (6792)	-184 to 77 (-120 to 25)	0.38	24.6hr / 687d
Jupiter 66 moons	5.2	88,846 (142,984)	-238 (-150) Cloud top	2.36	9.8hrs / 11.9 yrs.
Saturn 62 moons	9.5	74,897 (120,536)	-292 (-180) Cloud top	0.92	10.2 hrs / 29.5 yrs.
Uranus 29 moons	19.2	31,763 (51,118)	-346 (-210)	0.89	17.2 hrs / 84 yrs.
Neptune 13 moons	30	30,775 (49,528)	-364 (-220)	1.1	16.1 hrs / 164.8 yrs.
Pluto 6 moons	39.4	1,485 (2390)	-382 (-230)	0.04	6.4d / 248 yrs.
Asteroid Belt	2 to 3.5	Dust to 600 (1,000)	−300 to 15 (-190 to -10)	< 0.005	Varies widely
Kuiper Belt	30 to 50	up to 900 (1500)	Uncertain	< 0.005	Varies widely

TOUR OF THE EARTH

Earth's orbit is a nearly circular ellipse with an elongation of only 0.0167, but that means the nearest point to the sun (perihelion) is approximately 5 million miles closer than the farthest point (aphelion). This results in an energy difference of 6.9 percent. The Earth's axis always points toward Polaris, the pole star, so that as it travels around the sun, sometimes the Northern Hemisphere points more toward the sun and sometimes the Southern Hemisphere points more toward the sun, creating our seasons. Southern Hemisphere summers occur near perihelion (January 3), so they may be hotter. Similarly, winters in the Southern Hemisphere may be colder because they occur near aphelion (July 4). Greater ocean areas in the Southern Hemisphere help to mediate this effect.

High tides occur about every 12 hours on opposite sides of the Earth because of the rotation of the Earth under the gravitational influence of the sun and the moon which pull on the Earth, the oceans, and the atmosphere. During a month, the moon orbits the Earth every 27 days, so that when the moon and sun are on the same (new moon) or opposite (full moon) sides of the Earth, tides are greater due to combined gravitational effects. Tides also occur in the Earth and atmosphere, although less noticeably. Tides and wind circulate ocean waters and atmosphere to help cleanse and distribute waste, nutrients and useful gases like oxygen and carbon dioxide.

The Earth's interior is separated into crust, mantle, and core. The depth of the Earth's crust is, on average, 4 to 40 miles (6 to 64 km) thick in oceans and continents respectively. The crust is a thin layer on which lighter continental rock "floats" on top of heavier basaltic oceanic rock. In reality, the entire crust creeps across the surface of the Earth from spreading volcanic cracks where new crust is forming, eg, mid-ocean ridges, to subduction areas where crust is plunged back into the Earth's mantle to be recycled. The Earth's crust is divided into seven major tectonic plates and several minor ones. These plates can collide, causing new mountains to be pushed up, or be pulled apart forming rift valleys. The highest surface point is Mount Everest, at 8,848 meters (29,029 ft.), and the lowest point is the Challenger Deep in the Marianas Trench at 10,911 meters (35,797 ft.) below sea level. Even at that, compared to the bulk of the Earth, the surface is extremely smooth, with maximum variance at only plus or minus 0.08 percent.

The crust makes up only 1.6 percent of the Earth by volume. The mantle fills about 80 percent of the Earth's volume and consists of high temperature, high pressure molten or deformable rock. The mantle has convection currents from rising hotter and descending cooled material that drives the crustal movements and volcanism and can affect local gravity and magnetic fields. The core makes up about 16 percent of the earth by volume and consists of an outer liquid layer and an inner solid portion consisting mostly of iron with nickel. Note that the density rises from $2\text{-}3\text{g/cm}^3$ in the crust to $3.4\text{-}5.6\text{g/cm}^3$ in the mantle to $10\text{-}13\text{g/cm}^3$ in the outer and inner core, so that by weight, it would be more like an estimated 0.7 percent crust, 66 percent mantle and 33.3 percent core.

The rotation of the Earth creates a dynamo effect in the core that is responsible for Earth's magnetic field. The magnetic poles do not coincide with the axis poles around which the earth rotates. The magnetic poles wander so that the north magnetic pole presently resides about 5° south of the axis pole (true north) near northern Canada and is heading west/northwest at about 60 km (35 miles) per year. The north and south magnetic poles are not exactly opposite each other, so that the south magnetic pole resides about 30° north of the axis pole (true south) off the coast of Antarctica and is heading northwest at about 15 km (10 miles) per year. The magnetic field lines between the poles on which our compasses depend also wander and have eddies and excursions caused by local factors or interior circulation, so that the US and international geologic services must redraw them every few years to keep marine navigation charts current. Without the proper correction factors, navigation by compass would be very inaccurate. Fortunately, we now have GPS satellites that can be used to augment navigation by triangulation.

The magnetic poles can flip on geologic (long) time scales so that north becomes south, and have done so several times in the past. The last confirmed reversal was about 780,000 years ago, as seen in bands of reversed magnetism in solidified lava at the mid-ocean ridges. Reversals are thought to take tens or hundreds of years to develop. During these periods, the magnetic field drops to a few percent of present levels. This can open the Earth to deadly rays and particles from the solar wind and space that are normally deflected by the magnetosphere.

Deeper is hotter in Earth's crust due to heat from the mantle, radioactive decay, and pressure from gravity. At a few tens of meters, caves often are near a constant 50°F (10°C) due to air and water circulation and absence of direct solar heating, but in deep mines and natural caves the temperature rises to the point that life is unbearable without external cooling devices. On average the temperature increases at 25°C/km (124°F/mile) depth. At the Kola Peninsula experimental borehole[14], at a depth of 12,262 meters (40,230 ft. or 7.6 miles) the temperature rose to a whopping 180°C (356°F) although it was expected to only reach 100°C (212°F) in that location. Work was stopped because the drilling equipment could not stand any higher heat.

In the oceans, the pressure steadily increases with depth. Pressure increases by one atmosphere for each 10 meters (33 ft.) depth, so that at 200 meters, the pressure is 20 atmospheres. At the deepest part of the ocean[15] the pressure is about 1100 atmospheres. Light is reduced at greater depths so that below 200 meters, light is reduced to the point where photosynthesis is not possible. The oceans cover over 70 percent of the Earth, including 60 percent covered by waters deeper than 200 meters depth. The average depth is over 3,650 meters (11,975 ft.).

Temperature also declines with depth so that by about 1000 meters (3300 ft.) the average temperature is below 12°C (54°F). Average temperature at the bottom of the deep ocean is around 2°C (36°F), and at the deepest point in the oceans the temperature is about 1°C (34°F). The exceptions to this are the few volcanic regions such as hydrothermal vents in which the temperature of liquid[16] water may be as high as 450°C (842°F), supporting a rich community of life forms in the surrounding waters. Living creatures, from microbes[17] to crustaceans and fish, are present at all depths in the oceans.

The atmosphere is divided into zones called the troposphere, stratosphere, mesosphere, thermosphere and exosphere, but the temperature profile is a little more complicated than that of the interior of the earth or the oceans. Although, as a rule, higher is colder due to heating at the Earth's surface, it is not a steady gradient. At altitudes of commercial airlines, typically 10 to 15km (6 to 9 miles), the temperatures are in the range of -50°C to -60°C (-58°F to -76°F). However, temperatures rise from there to a maximum of around 0°C (32°F) in the stratosphere (50 km or 30 miles) due to heating by absorption of UV radiation in the ozone layer. Temperatures again dip down in the ionosphere (80 to 90 km or 50 to 56 miles), (which is in the mesosphere), to -90°C (-130°F). At the altitude of the thermosphere and upward, the air molecules are so thinned out that gas properties and temperatures as we know them are meaningless. While molecules, practically unimpeded by collisions with other molecules, may be moving so fast that their individual "temperatures" may reach several thousand degrees, their overall effect still results in extremely cold measured temperatures near -90°C (-130°F).

Average atmospheric pressure at sea level is 14.7 Psi (1.03 kg/cm)[18], but rapidly declines with altitude due to reduced density of air molecules and reduced depth of the air column. On the highest mountains the pressure, density and therefore the oxygen content reduction require supplemental oxygen. Airliners must be pressurized at most altitudes. In the same way that many deep ocean creatures require the crushing pressure of the depths to survive, complex life on the surface of the earth requires the pressure of the atmosphere to survive. We live at the bottom of a sea of atmosphere. Some birds migrate at altitudes that would suffocate us, and some bacteria are found as high as the Stratosphere, but as a general rule, we depend on the surface pressure, temperature and mixture of gases to keep us alive.

Only in this thin band of atmosphere and crust are conditions truly habitable for us. Although life exists in all climates and conditions, most complex life, from quadrupeds to humans requires a narrow range of temperatures, and plant life (crops) also require sufficient water and nutrients in the soil. The rest of the Earth, the solar system, and indeed the universe as a whole are completely hostile to us.

CHAPTER 7

TOUR OF THE INFINITELY TINY

FROM SUBATOMIC PARTICLES TO LIFE

At the base of all of this immense grandeur are atoms made up of negatively charged electrons orbiting nuclei composed of positively charged protons and electrically neutral neutrons. The atom is the smallest unit that still can be identified as a particular chemical element. If that was the whole picture, it would be relatively simple; however, there are even finer divisions and many more particles. Let me see if I can give you an overview without getting too deep or confusing. I apologize in advance for any tedious or confusing details.

According to the Standard Model of particle physics, protons and neutrons are actually composed of three smaller entities called quarks and are thus classified as baryons, while electrons are indivisible and are classified as leptons. The neutron consists of two down and one up quark held together by gluons, (a particle that carries force); the proton consists of two up and one down quark also held together by gluons. There are actually six types of quarks with the frivolous names of up, down, top, bottom, charm and strange, and six types of leptons called electron, muon, tau, electron neutrino, muon neutrino and tau neutrino particles.

There are also anti-particles for each of these, such as the positron, the antiparticle to the electron, antiproton, antineutron and muon antineutrino. Matter and antimatter, when combined, annihilate each other with the release of tremendous energy. Einstein's formula, $E = MC^2$, means the energy you can get from matter is the mass times the speed of light squared. That's a lot of energy for a tiny bit of mass and is the basis of the power of the atomic bomb, which splits atoms to release energy. It is also the basis of fusion energy in the sun, which combines four hydrogen atoms in a three-step process to form a helium atom with the release of essentially massless electron neutrinos and energy in the form of gamma rays from the matter/antimatter annihilation of a released positron by an electron.

In addition to these particles, there are particles that carry the forces that maintain and govern the behavior of atoms. There are three gauge bosons (called W boson, Z boson, and gluon) and the photon, as well as the Higgs boson that may have been found but has not been confirmed experimentally. The W and Z bosons carry the nuclear weak force holding the nucleus together. The gluons (of which there are eight varieties) carry the nuclear strong force that holds the quarks of neutrons and protons together, as noted earlier. The photon carries the electromagnetic force that holds the electrons to the nucleus.

The Higgs boson is theorized to be responsible for mass or gravity. If found, it could, potentially, lead to a "Theory of Everything," which includes the fourth force, gravity. At present, the picture of the weak nuclear, strong nuclear and electromagnetic forces, defined by the Standard Model of quantum physics, is incompatible with our picture of gravity by Einstein's Theory of Relativity. One or both of these theories may be wrong, or at least incomplete.

An electron can be bumped to a higher energy orbital within an atom by absorbing energy, and then it will fall back to a more stable, lower-energy orbital by emitting the energy as a photon. Only photons of a specific energy (frequency) can be absorbed by any one electron depending on its energy level. These distinct energies are called quanta as are the distinct orbital levels, thus the name "quantum physics." There is always a loss of energy in this process, so that the photon that is emitted when the electron falls back to its ground state is always at a lower energy than the photon that is absorbed. The absorption or emission of photons at distinct frequencies gives us a powerful tool to identify what elements are present. The family of absorption or emission lines in the spectra of stars is based on this principle and is used to identify their elemental content.

The mass of the neutron is 1.6749×10^{-24} grams. The mass of the proton is 1.6726×10^{-24} grams. The mass of the electron is 9.1094×10^{-27} grams. More simply, the electron is 1/1836th of the mass of the proton or 1/1839th of the neutron, so the atomic weight of an element is mostly due to the protons and neutrons in the nucleus. Sorry, there is no easy way to illustrate these tiny weights, but maybe it will help to put them in more common terms. To make one ounce, it would take 1.6926×10^{25} neutrons, or 1.6949×10^{25} protons or 3.1121×10^{27} electrons[1]. The elements from which all matter is composed consist of atoms with varying numbers of neutrons, protons, and electrons, usually in equal numbers except for isotopes having one or more neutrons less or more.

There are ninety-four naturally occurring elements and twenty-four confirmed heavier elements created artificially. The lightest element is hydrogen, with only one electron encircling one proton (atomic weight 1.00794 g/mole), then helium with two electrons encircling a nucleus of two protons and two neutrons (atomic weight 4.002602 g/mole). Some important elements include carbon with 6 electrons, protons and neutrons (atomic weight 12.0107 g/mole), nitrogen at 7 of each (atomic weight 14.0067 g/mole), and oxygen, at 8 of each (atomic weight 15.9994 g/mole). Plutonium-94, atomic weight approximately 242, is the heaviest confirmed naturally occurring element, and ununoctium-118, approximate atomic weight 294, is the heaviest confirmed synthetic element to date (ca. 2010).

A mole of a substance has mass in grams equal to the substance's molecular or atomic weight[2]. Most elements also have isotopes, usually containing added or reduced numbers of neutrons, naturally occurring as a small percentage of the total. Note that atomic weights listed are actually the averages of all isotopes in their naturally occurring ratios. Some of these isotopes are unstable and decay to the stable state radioactively at a specific unique rate, ie, half-life. One good example is carbon, with a standard atomic weight of approximately 12, but that can occur as carbon-13 and carbon-14, containing one and two extra neutrons respectively.

SOLAR FUSION OF HYDROGEN INTO HELIUM-4

In the core of the sun the temperature is so high that protons and electrons from hydrogen are essentially free and reach speeds high enough to fuse. (Temperature is a measure of the average speed of particles, the higher the faster.)

1. In step one, two protons fuse into a deuteron, made up of one proton and one neutron, by converting one proton into a neutron with the release of an almost massless electron neutrino and a positron (antiparticle of the electron). In the sea of fast and free electrons, the positron quickly finds and annihilates with an electron releasing tremendous energy in the form of gamma rays.
2. In step two, another proton fuses with each deuteron to form helium-3, containing two protons and one neutron.
3. In step three, two helium-3 nuclei fuse and form a helium-4 nucleus, containing two protons and two neutrons, with the release of two protons.

The total reaction to produce helium-4 uses two electrons and six protons, releasing two protons at the end, and by converting two protons to two neutrons releasing two positrons and two neutrinos.

> Additional gamma rays released by binding energy are produced with each fusion event in the chain.
>
> The two positrons annihilate with two electrons releasing gamma rays, which may take thousands of years to reach the surface of the sun through countless interactions with matter. Losing energy with each interaction, they appear mostly as UV, visible and infrared light at the surface.
>
> The two electron neutrinos, which rarely interact with other matter, leave the sun immediately, bombarding the earth with an estimated trillion particles per meter each second. Most of these neutrinos pass harmlessly through the earth and continue on their way. The sun is estimated to convert 600 million metric tons per second of hydrogen into helium-4, annihilating 4.3 million metric tons per second into pure energy. Only about a third of the expected electron neutrinos are detected on earth, so physicists have theorized that neutrinos oscillate between the three forms, electron, muon and tau neutrinos. The other two forms require different types of detectors. So far, muon and tau neutrinos have only been detected in particle accelerator experiments, but oscillation from electron neutrino to muon neutrino may recently have been detected from space.

Carbon-14 is created when thermal neutrons, formed by cosmic rays, hit Nitrogen-14 in the atmosphere[3]. Carbon-14 decays with a half-life of 5,730 ± 40 years back into stable nitrogen-14. The term "half-life" means that half of the mass is radioactively converted into its more stable form in that time. Plants take in the carbon-14 as carbon dioxide and animals eat plants or other animals that ate plants, so their bodies approximate the level of carbon-14 in the atmosphere (or ocean). Accumulation stops when they die. The convenient half-life length makes it a good marker to date ancient organic artifacts up to 60,000 years old[4]. Unlike carbon-14, carbon-13 is a relatively stable isotope that is present at approximately 1 percent in all carbon on Earth. Isotopes have similar but different properties from the standard state due to the difference in weight.

The diameter of the smallest atom, helium, enclosing the nucleus and two electrons orbiting it, is approx. 64×10^{-12} meters (64 picometers)[5]. The diameter of the helium nucleus, containing 2 neutrons and 2 protons, is about 3.4×10^{-15} (3.4 femtometers). So the diameter of this atom is 18,692 times the diameter of the nucleus, or over 6.5×10^{12} (6.5 trillion) times the volume of the nucleus. This means that the helium atom is made up of mostly empty space between the nucleus and the electrons. This is true within and between all atoms and molecules.

SPACE, THE FINAL FRONTIER

What appears to be solid matter on the larger scale, is really atoms containing, and contained in, mostly empty space interacting with other atoms electronically in an ordered system. It is possible for a photon or other particle to pass through without ever hitting a single atom or subatomic particle. This is precisely what happens when X-rays pass through the body, only leaving shadows on the film where some of the rays encounter denser material like bone. This is why the trillion neutrinos per meter per second, formed and emitted by the sun, can pass through the entire earth all of the time, rarely hitting any other particle.

So if the universe is composed of 700,000 trillion trillion times as much space as matter, and even the matter, within and between atoms, is mostly empty space, it could be said that statistically the amount of matter in the universe is negligible or practically nonexistent.

MOLECULES – THE BASICS

Atoms of the elements combine into molecules in pairs, clusters, branched and unbranched chains, rings, et cetera, and form solids by arranging themselves in lattices (crystals like diamond), sheets (like graphite) and glasses (poorly ordered conglomerations), or as liquids and gases, which are less ordered depending on the temperature. Everything we see, feel or interact with is composed of chemicals made of elements or molecules, all made of atoms. Everything that physically exists is composed of chemicals, so, if a commercial product is advertised as "chemical-free" they are lying. They may mean there are no synthetic (manmade) chemicals added, but the product cannot be chemical-free if it exists.

Atmospheric gases are small molecules in which usually two or more atoms or elements combine, such as N_2, O_2, Ar, CO_2, H_2. Air contains 78 percent nitrogen, 21 percent oxygen, 0.9 percent argon and 0.038 percent, carbon dioxide, but hydrogen is not a significant component of air unless you count water vapor, average 1 percent, dissolved in the air. Note that argon (Ar) is not paired because it is a noble gas, like helium and neon, that does not combine readily because their electron orbitals are complete, ie, have just the right number of electrons to complete each orbital making them stable. Combine hydrogen and oxygen (two gases) to get H_2O, water, which can be a liquid, solid or gas depending on the temperature and pressure.

When UV rays from the sun or lightning split water vapor, hydrogen and oxygen[82] are released as charged atoms called "ions." These are very chemically active and quickly combine with other atoms to form, for example, ozone, a very chemically unstable compound made up of three oxygen atoms. UV rays from the sun continually produce ozone by splitting and recombining oxygen atoms in the upper layers of the atmosphere by this process, and the ozone helps to shield the earth from harmful UV rays.

When rain water combines with carbon dioxide, carbonic acid is formed, ($CO_2 + H_2O = H_2CO_3$ or as ions $HCO_3^- + H^+$). Rain water is usually slightly acidic at pH 5 or so (not neutral pH 7) because of these reactions. As a matter of fact, any water left sitting open to the air will assume a similar acidic pH by absorbing carbon dioxide from the air unless buffered by other molecules. Buffering is a process where two components "fight" each other to maintain a particular pH balance. Fortunately, most systems containing living organisms contain ions that are very good buffers such as phosphates and carbonates (PO_4^{-3} and CO_3^{-2}). Due to this buffering effect, oceans average pH above 8 (slightly alkaline) in spite of the fact that 50 percent of CO_2 produced is absorbed by the oceans.

The most abundant element on earth is oxygen at 46 percent of crust, 79 percent of ocean water, 65 percent of the human body and is only surpassed by Iron in the total composition of the earth at 30 percent, with iron (mostly core) estimated at 35 percent. Living things are based on large carbon-based molecules containing oxygen, hydrogen and nitrogen with a smattering of other elements such as phosphorus, sulfur and calcium.

Nonliving things are mostly composed of other elements such as silicon, magnesium, aluminum and iron, usually in complex multi-element combinations with elements such as oxygen and sulfur. Nonliving (inorganic) compounds can contain ions such as carbonates and phosphates which may have been originally derived from living things (organic). For instance, limestone is composed largely of calcium and magnesium carbonates and oxides that may have originally derived from the shells and skeletons of microscopic life laid down over millennia as sediments in shallow seas.

WATER, THE ELIXIR OF LIFE

Water is truly a miraculous liquid that allows and facilitates life. Most other molecules in the same low molecular weight range are gases at atmospheric pressures and temperatures. Water remains in the liquid state over a wide range of temperatures and pressures due to hydrogen bonding. Hydrogen bonding is where the positively charged hydrogen atoms in one molecule are bonded to the electrons of atoms in other molecules. Water, with two positive hydrogen atoms bound to one negative oxygen atom, in a wide V shape, forms agglomerations of oriented molecules that are harder to break apart than most liquid arrangements.

Physical properties that are affected include boiling point, freezing point, higher viscosity, higher surface tension and lower vapor pressure.

COMPARISON OF LIQUID PROPERTIES OF WATER TO METHANOL

	Viscosity	Surface Tension	Vapor Pressure	Dielectric Constant at 25C	Flash Point
Water H_2O	1.002 centipoise	73 dynes /cm	17.5 mm	78.54	none
Methanol CH_3OH	0.593 centipoise	22.6 dynes / cm	92 mm	32.63	12.2C (54F)

Due to the hydrogen bonding, water also has a couple of other properties that make it more favorable to life. Once the boiling point of water is reached using 1 calorie/gram/degree, it takes an additional 540 calories/gram to produce steam, ie, boil. Similarly, it takes an additional loss of 80 calories/gram to freeze water into ice at the freezing point. Note that these high heats of condensation and fusion can produce superheated or super cooled liquid water. Tiny droplets of atmospheric water can remain liquid at extremely low temperatures until nucleated by a particle such as ash, dust, salt, meteoric debris or even a bacterium, at which time they instantly freeze and attract other water droplets, which is the first phase of forming rain or snow. The purity of evaporated water and the low pressure at the altitude of clouds contribute to this. The urban legend that a cup of water heated in the microwave oven can violently erupt spontaneously is based on superheating of extremely pure (distilled) water. Introduction of any nucleating substance such as tea or a spoon can cause instant boiling. This is rarely seen because most water contains enough minerals to prevent or reduce superheating.

Along with the low vapor pressure and high surface tension inhibiting evaporation, this means that large bodies of liquid water are possible. Unlike most other liquids that continue to shrink as they freeze, water expands by about 10 percent. This means that ice floats due to reduced density; otherwise the oceans would freeze solid from the bottom up. Water has a high heat capacity that makes it an excellent heat sink to regulate atmospheric temperatures, both as a liquid and as vapor in clouds. Clouds also reflect a significant amount of heat from the sun back into space. Water strongly absorbs infrared radiation (heat) from the sun, but it is transparent to UV and visible light. Attenuation of these wavelengths at depth in the oceans is due to salts and organics dissolved or suspended in it. Due to its polar nature and ability to split apart into H_3O^+ and OH^- ions, it has a high dielectric constant and readily dissolves most salts and many other molecules.

See table below for comparison of other properties of water to some typical low molecular weight molecules. Most are gases, but note lithium hydride and boron nitride, which are solids.

PROPERTIES OF OTHER LIGHT MOLECULES COMPARED TO WATER

NOTE: Standard Atmospheric Pressure and Temperature are assumed unless noted otherwise (760mm Mercury and 20C)

Compound	Formula	Molecular Weight (rounded)	State	Boiling Point	Freezing Point	Soluble in Water	Other Properties
Water	H_2O	18	Liquid	100C	0C		Heat of condensation, 540C; Heat of Fusion, 80C.
Methane	CH_4	16	Gas	-161.6C	-182.5C	Slightly	Toxic; Flammable
Carbon Monoxide	CO	28	Gas	-190C	-207C	Slightly	Toxic
Carbon Dioxide	CO_2	44	Gas	NA	-78.5 Sublime[7]	Yes	Animals exhale it; Plants use it; Dry Ice.
Acetylene	C_2H_2	26	Gas	-84C	-81.8 at 890 mm pressure	Slightly	Flammable
Methanol	CH_3OH	32	Liquid	64.5C	-97.8C	Yes	Toxic

Lithium Hydride	LiH	8	Solid	Decomposes	680C	Yes; Decomposes	Unstable in the presence of water
Boron Nitride	BN	25	Solid	NA	3000C Sublime		hygroscopic (attracts water)
Hydrochloric Acid	HCl	36	Gas[8]	-85C	114C	Yes	Corrosive
Hydrofluoric Acid	HF	20	Gas	19.5C	-83C	Yes	Corrosive
Hydrogen Sulfide	H_2S	34	Gas	-60.2C	-83.8C	Yes	Toxic
Nitrogen	N_2	28	Gas	-195.5C	-210C	Slightly	80% of air; toxic at high levels
Oxygen	O_2	32	Gas	-183C	-218C	Yes	20% of air; corrosive at high levels; promotes combustion.

Water is one of the most abundant molecules on the planet. Oceans and other bodies of water cover over 70 percent of Earth's surface; about 96 percent of Earth's water is salt water in oceans and over 3 percent is in glaciers or ground water. Less than 0.5 percent is in freshwater rivers, lakes and streams. Water is also contained in living things and as water of crystallization in many rocks and minerals. For example, the familiar blue copper sulfate crystal is composed of hydrates containing 5 water molecules in the crystal lattice called copper sulfate pentahydrate. Chemically it is written as $CuSO_4.5H_2O$, or more precisely $[Cu(H_2O)_4]SO_4.H_2O$ reflecting the actual structure.

The water cycle consists of evaporation and plant transpiration into water vapor, clouds and fog (ground clouds) followed by condensation and precipitation. In most areas of the earth, dew or frost is deposited on most nights of the year when temperatures fall below the dew point – the temperature where the amount of water vapor in the air reaches the saturation point and coalesces on surfaces into droplets – even in very arid places. Plant life moderates temperature swings between night and day, but the relatively high humidity near plants results in formation of dew on most nights. In deserts, temperatures swing widely from scorching hot to freezing cold, so that even low humidity can reach saturation and become dew, frost or fog at night.

Although oxygen is optional or undesirable for some bacteria, water is required for all life. The interior of cells, including organelles, DNA and proteins, and multicellular organisms are bathed in water solutions that facilitate life's processes.

OF MOLECULES AND LIFE

The size of molecules ranges from approximately 3×10^{-10} meters (30 billionths of a meter) for H_2 to several meters for DNA, the inheritable genetic material that facilitates life processes inside each cell. Each molecule of human DNA is about 3 meters long, and contains approximately 6 billion base pairs [9] made up of about 400 billion atoms. Human DNA, contrary to what you may think, is not the longest or the most complex. It forms 46 chromosomes[10] composed of DNA tightly wound with proteins. Compare that to 48 for chimpanzee, gorilla, orangutan, beaver and deer mouse, 78 for dog, wolf, hare, chicken and dove, 8 for fruit fly, 6 for Mosquito, 20 for corn, and a whopping 1,260 for adders-tongue fern (highest number known). As a matter of fact, many "lower" life forms have the higher numbers of chromosomes. Clearly, the number of chromosomes does not correlate to advanced species.

Proteins are the work horses of life as structural units, molecular machines such as flagella and enzymes that perform specific metabolic functions. Proteins are composed of strings of hundreds or thousands of amino acids in very specific order to perform specialized functions. Each protein is internally linked and folded on more than one level to form molecular structures and machines, often in concert with other proteins. Among the molecular machines, enzymes act to cut, combine, hold and reorient other molecules that perform work such as metabolism. Each of the 20 amino acids[11] used to build proteins is encoded by a sequence of three specific base pairs of DNA. Genes contain sections of DNA made of a series of these triplets that encode for specific proteins and other non-coding DNA sections. RNA first copies the specific section of DNA, and then takes this information to special organelles called ribosomes that then read the RNA to build proteins, one amino acid at a time. This is a simplified picture of processes that are much more complex involving cutting, splicing and rearranging the mRNA (m for messenger).

DNA occurs as two copies, one from each parent in normal cells, and is packaged in very complex ways with proteins that can function as either structural supports or that facilitate gene regulation and expression. At the first level of organization, DNA is wound around roughly spherical complexes of eight special proteins called histones, often referred to as a "pearls on a string" arrangement, (more like "spools on a string"). Genes are most accessible in this form. As the DNA prepares for cell division, after making a copy of each strand, each copy is supercoiled into a condensed rope or "fiber" form. These supercoiled, tightly wound sections are then organized into stacks of loops around long scaffolding proteins, then into more condensed form with additional structural proteins to form chromosomes. The chromosomes are then pulled apart and the cell divides into two identical daughter cells. Please keep in mind that this is a necessarily simplified picture of a much more complex process. Also note that each of these molecules is not "alive" themselves. All of these processes are chemical in nature and are facilitated by a host of other molecules.

Regardless of the size or complexity of the organism, these facts are true for all living things except viruses. Only the details of the process vary between prokaryotes and eukaryotes. Prokaryotes, consisting of bacteria and archaea, are without a separate nucleus and contain circular DNA, whereas eukaryotes, including everything above bacteria, from yeast and amoebae to humans, are composed of cells that have a nucleus which contains linear DNA. Both also have separate short sections of DNA in other structures such as ribosomes[12], mitochondria[13] and plastids[14].

Viruses consist of DNA or RNA enclosed in a protein shell and do not have the machinery for metabolism or to replicate by themselves. They inject their DNA into other cells and hijack their replicating machinery to make many copies of themselves that are then released through cell disruption. Most biologists believe that viruses are not alive, but they definitely use their own machinery to invade and inject their DNA into other cells. So, even though much of the viruses' life cycle occurs inside other cells, they cannot be said to be dead either. They are more like parasitic spores. Conventional thinking about viruses has been put into question by recent discoveries of viruses as large as 1 micron (micrometers or millionths of a meter); which is 100 times as large as is typical, and is in the size range of some bacteria, with DNA large enough to rival both bacteria and some simple eukaryotes[15].

LIVING THINGS

Typical viruses range in size from 20 to 400 nm (nanometers or billionths of a meter). Bacteria range from 0.1 to 750 microns (micrometers or millionths of a meter) although most bacteria are in the 5 micron range. Organisms that contain a cell nucleus (eukaryotes) contain one or more cells that range from about 10 microns to several feet for nerve cells to multicellular organisms of several hundred tons for things like redwoods. The largest single cell is the ostrich egg.

Life on Earth is practically ubiquitous[16]. Bacteria and archaea[17] microbes are found in almost all known environments from the stratosphere to deep crust, from frozen arctic regions to hot deep-sea thermal vents, from deserts to oil wells to jet fuel tanks to radioactive nuclear wastes to very hot, very acid or very alkaline or very salty environments. Microorganisms have been found 25 miles (40 km) high in the atmosphere and down to 3 miles below the Earth's surface with no end in sight. The deeper you go into the Earth's crust, the hotter it becomes, and so there must be some depth where life ends. Microbes can survive at temperatures up to 250°F (122° C) in the high pressures of oceans, and are found in the ocean down to depths of 6 miles (10 km) in the Marianas Trench, the deepest in the world.

They make up a significant portion of the biosphere and may exceed that of all other life forms in total mass. They even live on us, inhabiting places like our skin and gut, and are estimated to contribute significantly to our weight. They inhabit the guts of all animals to help them digest their food into a form that can be absorbed by cells, and can produce vitamins the animal itself cannot make on its own. For example, termites would not be able to digest the complex cellulosic fibers of wood without the symbiotic[18] archaea microbes in their gut that supply them with all the nutrients they need.

The earth is believed to be 4.5 billion years old, and possible signs of microbes are found in rock strata as far back as 3.8 billion year ago. The earliest fossils discovered of multicellular soft-bodied life occur at about 6 million year ago[19].

The earliest fossils of hard-bodied or shelled life occur at 5.4 million years ago in the so-called Cambrian Explosion, which contained all existing body types. Single celled eukaryotes consist of a variety of protozoans, diatoms, algae, yeasts and fungi. The simplest multicellular organisms are colonies with little or no differentiation or specialized cells. Beyond that, are true multicellular organisms with specialized cells performing specific functions to benefit the entire organism. Plankton, on which all of ocean life is based, consists mostly of microscopic single celled and multicellular organisms.

Almost all of life on Earth ultimately depends on photosynthesis, which uses the energy of sunlight to turn water and carbon dioxide into sugars or starches, releasing oxygen. The only exceptions are those few that derive energy from other sources deep in the crust or oceans. There is an elaborate balancing act whereby plants produce food and oxygen from sunlight, carbon dioxide and water, while animals eat the food to build other molecules and derive energy while releasing carbon dioxide and organic wastes that feed the plants.

The most primitive photosynthetic organisms are specialized cyanobacteria, also called "blue-green algae," that were present very early in Earth's history. Although they use a primitive form of photosynthesis, it is similar, and indeed most of it is a part of the photosynthetic mechanism used by plants today. Cyanobacteria use a pigment called phycocyanine, instead of chlorophyll used by green bacteria and higher plants, to collect energy from sunlight for photosynthesis, but use the energy in the same way. Some scientists think the chloroplasts in plants originated as captured cyanobacteria or similar microbes forming endosymbionts[20]. Both have the same double stacked internal membrane layers (lamella), circular DNA and ribosomes that perform similar functions. Cyanobacteria are also symbionts of other organisms such as coral.

It is thought that much of the oxygen in the atmosphere was first produced by early cyanobacteria under anaerobic (excluding oxygen) conditions[21] releasing oxygen as a waste material that was toxic to them. Aerobic (using oxygen) microbes are thought to have developed later to use this waste product. An atmosphere with low or no oxygen would have lacked an ozone layer so that the surface would have been bathed in deadly UV light. Any life forms at that time would have needed shielding such as deep water or soil. Additionally, before photosynthesis, all life would have had to depend on energy sources other than sunlight such as hydrogen sulfide. Life may have begun either deep inside the earth or at the bottom of deep oceans in locations such as hydrothermal vents that supplied these other energy sources. After there was sufficient oxygen to form a suitable ozone layer, photosynthetic bacteria could then utilize sunlight and carbon dioxide directly from the atmosphere to produce more oxygen and reduce initial carbon dioxide levels.

At each stage, from anaerobic to aerobic to photosynthetic prokaryotes to single celled eukaryotes to colonies to multicellular organisms, required great leaps in complexity and functions. Then more adaptation was required to go from individuals to interdependent colonies such as ants, from aquatic to amphibian to land-based forms until every niche is thoroughly filled with vast communities of interacting organisms called ecosystems. These changes are thought to be the result of evolution.

PART 3

EVOLUTION AS THE MODEL OF PROGRESSIVE THOUGHT

CHAPTER 8

EVOLUTION: SETTING THE STAGE

The decline of the Holy Roman Empire in the fifth century was followed by a period of absolute power and control by monarchs and the Catholic Church known later as the Middle Ages or Dark Ages, so called by those in succeeding generations who wanted to believe theirs was a more "enlightened" age. Starting with the Italian Renaissance, in the fourteenth century and continuing through the Age of Reason and the Enlightenment of the sixteenth through the nineteenth centuries, the Western world was in a state of constant turmoil and social change.

The invention of the printing press by Johann Gutenberg in 1450 made mass printings and translations of the Bible and other books possible so that they became accessible outside of elite circles. This furthered the Protestant Reformation, which had begun in the fourteenth century with John Wycliffe in England and Czech priest Jan Hus, who was burned at the stake in 1415. These departures from Catholic tradition were followed by other reformers, including Martin Luther, a Catholic priest who posted his Ninety-five Theses in Wittenberg in 1517 condemning church abuses.

The medieval Catholic Church, with its domination by rich and powerful men, had become more of a political and military force than a representation of Christ's love and compassion as reflected by the apostles and the early church. This is not to say that the Church did not do many good things, as exemplified by church run universities, hospitals, orphanages, observatories, libraries and other repositories of ancient and new knowledge.

Forced conversions, persecution of heretics, ie, any form of Christianity,[1] science or philosophy not sanctioned by the Church, excommunications, torture, and death by burning and other means were all practiced at times by the Church. The reader should note that excommunication to the devoted Catholic of the time meant he had no chance of salvation and was doomed to hellfire forever, as opposed to Protestant belief that only God could ever decide that. What a freeing concept! Imprisonment, torture, and even death were preferable to excommunication. Similarly, monarchies had absolute power over the people and could, almost at will, have anyone stripped of his position, his property, his freedom, or his life. Only the Church had any power over monarchies.

Assisted by the rise of Protestantism, the alienation of the people by abuses and domination of the Catholic Church and the monarchy eventually led to limitations or overthrow of monarchies and the rise of various types of social philosophies and experimentation within and outside the various churches. In this vacuum, many experimental philosophies were espoused, some of which were irreligious or openly hostile to religion. Some went so far as to throw out Christianity altogether. The Directory, set up in France after the French Revolution, is probably the best, though later, example of such an extreme view. For a period of time under this regime in the eighteenth century, religion, particularly Christianity, was actually outlawed in France, which led to persecution of Christians and Jews.

The Italian Renaissance, beginning in the fourteenth century, marked a return to classical thinking of the Greek philosophers such as Plato and Aristotle. This influenced the formation of humanist and materialist philosophies, which became popular among intellectuals. Materialist philosophy states that there is nothing beyond the material world that we can see and touch. Humanists[2] believed that man was basically good and was only corrupted by society. Materialism[3] said "down with God," and humanism said "up with man."

The goodness and eventual perfection of man and society was/is an important part of the basic philosophy of Progressivism and was/is practiced by communists, socialists and their ilk. It is, in my view, a misinterpretation of human nature and a false belief in our ability to change it. This is magical thinking because it is contrary to experience. The unchanging nature of man is the reason that both Shakespearian and Greek plays still have relevance today. The circumstances and society are different, but the human reactions are the same. The Christian viewpoint was/is that man himself is imperfect and cannot be perfected by human endeavor, no matter how noble. If society is faulty, it is because imperfect man is its author.

Note that the word *progressive* has been corrupted from its original meaning. The Progressive philosophy originally meant that progress was possible through work, learning, and inspiration, built on the works of others progressively. This is contrary to today's interpretation of Progressivism as an inevitable quality of the universe, moving naturally from simpler to more complex and from imperfection to perfection and utopia, which is again magical thinking.

In the nineteenth century, Socialist thought was dominated by the "man good, society bad" humanist philosophy of Rousseau, and was guided by the "scarcity and struggle" philosophy of economist Thomas Malthus.[4] The dominant theme of the day was social progress of the "noble savage" toward ultimate social perfection, once he was freed from the tyranny of governments and religion. This utopian dream was a perfect philosophy for radicals who wanted to overthrow, rather than reform, what they perceived to be the corrupt and corrupting society of the time. Many social experiments were carried out in which utopian socialist communities were formed, lived and ultimately failed.

A very early forerunner of this was the Plymouth Colony, led by William Bradford in the seventeenth century. From the beginning they tried a form of communal living wherein all production was shared equally by everyone. When it became obvious that people would not produce well unless they were rewarded in proportion to their labors, this philosophy was quickly rejected, and was replaced by private ownership and free enterprise that quickly produced more prosperity and created wealth. Nineteenth-century examples of these utopian Socialist experiments include New Harmony, Indiana, Brook Farm, Massachusetts, and North American Phalanx, New Jersey.

SOCIALISM/COMMUNISM

Socialist thought began to be accepted in the seventeenth century (or even earlier) and flourished in the nineteenth and early twentieth centuries. Socialism is actually a kind of social Darwinism or social engineering. It is based on a misunderstanding of human nature and a belief that man's very nature could be molded and improved. This was based on the belief that the world is naturally progressive and everything, including human nature, is being continually improved throughout time.

The socialist dream sounds wonderful: everyone working for the common good and no one going without. Unfortunately, this belief has proven again and again to be wrong. Human nature is basically self-centered, and in general man is generous and altruistic only after personal needs and desires are met. Humans are motivated by a focus on me first, then spouse and children, then extended family, then friends, then local tribe, and only then extending to local and state authority, to country and to the greater global society last.

Man naturally is very compassionate and generous toward those in need of charity, but only after his basic needs are met. Socialism requires that man's focus be on the state (or society as a whole), while putting himself and his own interests last. This is the exact opposite of man's true, unchanging nature. Habits and attitudes can be taught to a certain degree, but it has been demonstrated many times that man's basic selfish and imperfect nature cannot be changed.

As the Plymouth Colony learned, without personal rewards for his achievements, a person's motivation to produce is reduced or eliminated along with most of his creativity and efficiency. At the same time, his selfishness, envy, resentment and deceit grow as a result of perceived inequities. In such a society, the lazy person who hardly contributes at all gets as much as the hard working person who produces most of what is shared.

In labor unions where all members are rewarded equally whether they are cracker-jack contributors or space-filling dead wood, resentment is rampant and efficiency and productivity suffer. Such unions discourage excellence and encourage minimal or status quo contributions. In the absence of an overarching internally motivated altruism, socialist societies must be tightly and thoroughly controlled by the state, ultimately resulting in totalitarian dictatorships or at best dictatorial bodies of an elite class in order to force people to behave as is required to maintain the society.

Unfortunately, socialism /communism also leads to moral degradation wherein cheating, lying and other forms of deceit are used to gain perceived or actual basic needs or an advantage over others. A prime example is the old Soviet Union, where morals and ethics have suffered greatly from real or perceived deprivations. As a general rule, needy is greedy. Everyone may be equal, but everyone, except the elite, is poorer for it.

Essentially we are back to monarchies and privileged gentry oppressing serfs or slaves "for their own good." So much for equality as espoused by socialism, communism and their ilk. It is a very old, very bad idea that results in a return to old oppressions and a loss of basic freedoms and inalienable human rights "endowed by our Creator."

But wait, what about the utopian dream? Karl Marx presented his philosophy as a series of steps where, through the principles of dialectical materialism,[5] society progresses from original oppression by the bourgeois[6] under capitalism[7] through struggle to a dictatorship of the proletariat[8] and then to a utopian state. Unfortunately, it never goes beyond the dictatorship stage, because the utopian dream is totally unrealistic, unworkable, and unsustainable in the real world due to the inherent and unchangeable nature of man and to reality in general. Karl Marx never explained how the society would take that final step from dictatorship to utopia. People in power want to stay in power. It is totally unrealistic to expect them to voluntarily give that up.

Even if utopia were attained, how would the utopian society be organized and maintained without essentially robotic altruism to the society[9] by every individual and (again) strict control from the top to keep it all going? Like monarchies and dictatorships it is still all about control by an elite group. In recent fiction, *Star Trek* is a model of a utopian society. Poverty has been eliminated and altruism is the norm. No one is envious or resentful of others' successes, and everyone gladly obeys orders from a wise and benign leader toward a common goal. However, the real world is more like *Babylon 5*, with all of its intrigues, envy, resentment, prejudices, hatreds, and inequities. Man's nature cannot be denied, and control through coercion and rewards is necessary for even a utopian society to function. Heaven on Earth is impossible as long as imperfect people are involved.

Unfortunately, even today there are those who would throw away their freedom, in the form of excessive regulation and government control, to gain a (false) sense of security under the control of a supposedly wiser elite. It seems there are some who are uncomfortable with freedom, with all its risks and opportunities, and who desire a nice safe cage. (Some intellectuals and elitists who espouse the Socialist philosophy assume that they will be among the elite and are only uncomfortable that others are not controlled. However, most of these people will end up being the controlled, not the controllers.)

> They, who can give up essential liberty to obtain a little temporary safety, deserve neither liberty nor safety."
>
> *- Benjamin Franklin*

Christianity is built on the value and importance of the individual whereby everyone benefits freely as a result of freely practiced moral values such as duty, honor, charity, respect and equality of opportunity, not outcome, which is unrealistic.[10] Socialism, in its many forms, does not value the individual but rather sees people as groups that should (be compelled to) work for the greater good of the whole, regardless of whether it is good for any one individual. These two philosophies are diametrically opposed. Socialism can only succeed if Christianity is either eliminated or tightly controlled as a purely social ritual. That is why socialism and atheism are such good partners, and behind socialism is the ever-present Progressivism.

The belief in a naturally progressive universe says that everything from the universe to molecules is evolving toward perfection, with no room for absolutes, not even moral ones. If the entire universe is believed to be naturally progressive, then there is no need for a God to have caused or influenced it. It is its own reason for being. To those who espouse atheism or Socialism in its various forms, Progressivism is what gives meaning to life and their cause, essentially replacing

God. It gives them a purpose and a satisfaction in furthering that assumed natural progress. That is why it has such a strong hold on its believers, especially those who wish to engineer a Socialist utopia or stop evolution in its tracks to save the planet.

The progressive universe itself becomes their de facto god, and social change toward a dreamed-of perfect utopian paradise becomes the goal and their purpose in life. Since no cultural or social system has ever achieved the perfection envisioned, to the Progressive the present system, whatever it may be, must be changed to further that perceived progress. This makes it a perfect philosophy for young radicals who wish things were better but lack the life experiences to see the broader picture or the unintended consequences of rampant social change.

However, remember the maxim: "all progress is change, but all change is not progress." That is, unless you believe that progress is inevitable as do the Progressives. But progress requires work while regress is the natural state of things. A boulder perched on the edge of a cliff, given enough time and erosion, will naturally roll down (regress) by necessity, but pushing it back up to the top (progress) requires work. Progress is not a natural thing; it must force its way against the regressive nature of the universe. The Second Law of Thermodynamics states that entropy always increases – that chaos or disorder always increases and usable energy always decreases. This is the opposite of the Progressive philosophy. Dust, death, and decay are natural results of the real world.

If the universe is naturally progressive, then everything must be interpreted as progressing or evolving toward perfection, whether it is molecules, life, Earth, stars, galaxies, or the universe. The fixed laws and values of basic physics, such as the force of gravity or the mass of the proton, are a great mystery to those who reject all absolutes in favor of universal progress. These values are under constant attack by theorists using deductive reasoning, ie, pure reason, rather than inductive reasoning based on reality, experiments and observation.

This is the case of cosmology and particle physics today. They start with an a priori[11] hypothesis, based on assumptions about how they believe the universe must behave, and produce complex mathematical equations to model an imagined perfectly symmetrical, homogeneous and beautifully progressive universe. In areas where reality conflicts with the theory that is based on pure mathematics, the facts are either ignored as anomalies, reinterpreted to make them fit or new layers of complexity are added to their calculations. Never is the theory questioned.

Why all the expounding on Progressivism and its partner Socialism, with its unrealistic view of human nature, and ultimately its tragic results? First of all, it is a perspective on the pseudoscientific theories discussed in this book that are all about control of thought by an elite class of experts who are not to be questioned. Progressivism and Socialism have influenced or control the foundations of most of modern science and academia today. That does not mean that real scientific achievement is not valid or does not advance our knowledge of our world; it means that real results are often interpreted to fit the prevailing Progressive paradigm.

For instance, if DNA of similar organisms is less different than dissimilar ones, which is expected if DNA determines form, it is not acceptable to just state the known facts and note the similarities and differences. The very real data must be fitted into the evolution paradigm by concluding that similar DNA means that they must have evolved from a common ancestor. While this may or may not be true, it is far from proven. It is a leap of faith and a philosophy based on existing paradigms.

Secondly, Darwinism and eugenics[12] specifically have been used as tools and extensions of socialist philosophy throughout its history. Pre-Marx Progressive Socialist thought itself nurtured Darwinism. Darwin's theory of evolution by survival of the fittest (class struggle in socialist parlance) through natural selection arose amidst this nineteenth century pre-Marxist Socialist-Progressive era. In the context of the prevailing philosophies, this meant to the materialists and humanists that once and for all religion could be eliminated. It seemed to confirm their social ideas that the world was naturally progressive and did not need any outside forces to bring it about. Using the theory of evolution, religion could be replaced by materialism, humanism, and Socialism as the new religion of the people. According to some accounts, Karl Marx actually wanted to dedicate his book *Das Kapital*, to Darwin but Darwin's wife objected.

DARWINISM

Darwin's circle of friends and mentors was largely composed of the intellectual elite of the day, many of whom embraced Progressivism, Socialism, atheism or agnosticism, and various other popular philosophies of the day. Darwin himself stated in some of his correspondence that one of his goals was to do away with religion.

After the initial presentation of a paper to the Royal Society, the philosophy, (theory), of evolution was published in the

> ... hardly see how anyone ought to wish Christianity to be true: for if so, the plain language of the text seems to show that the men who do not believe, and this would include my father, brother and almost all my best friends, will be everlastingly punished. And this is a damnable doctrine.
>
> - *Charles Darwin*

popular press, much like other popular philosophies of the day, not in scientific journals. Its arguments were more philosophical than scientific, offering little evidence other than similarity of forms between fossil and living animals, and observations that the existing forms were well suited to their functions, both of which had been widely accepted earlier. Contrary to popular accounts, from the beginning, many people in academia, the sciences, philosophy, and the clergy enthusiastically embraced the new philosophy of evolution. To the clergy, it was the means whereby God had created our world. To the anti-religion elite, it meant God could be replaced altogether, along with any inconvenient moral limitations.

Championed more like a political campaign than a scientific theory, after some early opposition by other scientists it became accepted by the dominant elite, so that scientists had to either adopt it or become obsolete. Any opposition was branded as ignorance or religious tyranny in heated debates where evolution proponents used a straw man[13] argument in which they presented Darwinian evolution versus creation ex nihilo.[14] Most people of the time recognized that changes had taken place, so that their logical arguments actually involved a lack of scientific evidence for the theory as presented. In some respects, that picture has prevailed to this day. It is this political tactic that has been repeated in other areas of science to promote new theories, to squash opposition to them, and for junior scientists to unseat senior scientists from positions of authority. That is why Progressivism and Darwinism, aka Evolution, is so important to later scientific philosophies and developments.

The theory of Evolution was based on the economic philosophy of Thomas Malthus, whose book, *An Essay on the Principles of Population* predicted that population would outgrow food supplies, resulting in starvation. Like Malthusian philosophy, the mechanism of evolution, survival of the fittest through natural selection, depended on competition for scarce resources as the basis of survival. In the introduction to the first edition of *On the Origin of Species*[15], Darwin explains evolution as, "this is the doctrine of Malthus applied to the whole animal and vegetable kingdoms."[16]

At the time, there were two opposing theories about the development of the Earth. One was catastrophism; the other was uniformitarianism. Catastrophism, supported by Georges Cuvier, the father of paleontology, proposed that the Earth had gone through repeated sudden upheavals. Uniformitarianism, promoted by Charles Lyell, geologist and friend of Darwin, proposed an Earth where no major changes had taken place except gradual modification over vast periods of time. Darwin had taken the first volume of Lyell's book, *Principles of Geology*, on his voyage around the world. Needless to say, Darwin favored Lyell's position. Later, Darwin accepted Lyell's theory as supporting his claims of gradual changes over vast periods of time. Cuvier, who had died before Darwin's time, had opposed uniformitarianism and the earlier evolutionary theories discussed in the next section. Evolution needed long eons of time for the proposed changes to take place, so uniformitarianism was the chosen philosophy that would facilitate it.

It is interesting to note that until the late twentieth century, uniformitarianism was the accepted dogma.[17] Today, as the best explanations for the fossil record and evolutionary changes, long periods of uniformity interspersed by brief catastrophic events of various sorts are favored. Thus, catastrophism is favored along with elements of uniformitarianism in the form of plate tectonics, formerly known as continental drift,[18] which had been rejected earlier. The renewed interest in catastrophism was fostered by the recognition of meteorite strikes and craters as a prehistoric reality that would fit past mass extinctions best.

I have witnessed the acceptance of catastrophism, meteorite craters, and plate tectonics, since the 1970s. When I first started my independent studies into science and Earth's mysteries, catastrophism, widespread meteorite craters, and continental drift were considered fringe theories. Serpent Mound, an earthwork by the prehistoric Adena or Fort Ancient culture in southern Ohio, is on the edge of an ancient four-mile-wide weathered meteorite crater. When I first visited Serpent Mound in the early 1980s, the visitor center still had the display claiming it was a crypto-volcanic crater. Although the strata were of dolomite and other sedimentary limestones with no hints of volcanic rock, the prevailing theory proposed an underground gas explosion caused by cryptic or hidden volcanism. Since that time, over 200 meteorite craters have been identified, most of them not readily recognizable due to weathering or other obscuring forces, including the one off the coast of Yucatan that is credited with the extinction of the dinosaurs at the end of the Cretaceous Period.

The re-acceptance of these theories is an example of how science should work. In science, inconsistencies in current theories are met by new data, and questions are answered by formation of new theories or acceptance of once rejected old ones. That is not to say that politics had nothing to do with it. On the contrary, the plate tectonics theory was pushed through in the popular press in the same way that Darwinian evolution was. Established geologists that did not immediately go along with the theory were publicly ridiculed and defamed in a way that could only be described as scandalous. It was a scientific revolution in geological circles.

EVOLUTION FROM THE BEGINNING

When Charles Darwin published *On the Origin of Species* in 1859, evolutionary theories had been around for a long time. The third-century BC Greek philosopher

Epicurus derived a form of evolutionary theory from Democretus' atomic theory. Lucretius, first-century BC Roman poet, proposed it as a logical necessity of naturalism in order to explain life arising from nature alone without divine intervention. It was resurrected in the Renaissance through the Enlightenment in a number of forms. (See the table.)

Charles Darwin had been introduced to evolutionary theories through his grandfather, Erasmus Darwin, a physician, inventor, and poet. Erasmus Darwin was a friend of William Wordsworth and Samuel Coleridge and their contemporaries who admired his poetry. Mary Shelly wrote *Frankenstein* after reading of his galvanic experiments on animals. He was an advocate of evolution by acquired characteristics, a theory later popularized by Jean-Baptiste Lamarck and which was still later discredited as having no viable mechanism.

Charles Darwin's family was wealthy, being associated with the Wedgwood fortune. Both sides of the family were Unitarian free thinkers, but the Wedgwood side leaned toward Anglican, at least socially. In this environment and later through his brother Erasmus' circle of friends, Charles was exposed to the intellectual elite of the day.

> Organic life beneath the shoreless waves
>
> Was born and nurs'd in ocean's pearly caves;
>
> First forms minute, unseen by spheric glass,
>
> Move on the mud, or pierce the watery mass;
>
> These, as successive generations bloom,
>
> New powers acquire and larger limbs assume;
>
> Whence countless groups of vegetation spring,
>
> And breathing realms of fin and feet and wing.
>
> *Erasmus Darwin, "The Temple of Nature"*

Charles' father Robert was a physician and wished for Charles and his older brother Erasmus to follow suit. The brothers attended medical training together at the University of Edinburgh. Erasmus graduated as a physician, but was retired on a pension at age 26 by his father because of his frail health. He spent the rest of his life entertaining the intellectual elite. Charles did badly in medicine, probably because his interests lay elsewhere. While there, he studied with naturalist Robert Edmund Grant, who was a proponent of Lamarck's acquired traits evolutionary theory and homology, a belief that similar form meant common ancestry. He learned stratigraphic geology from Robert Jameson and also studied plant classification and taxidermy.

Later, his father sent him to Christ's College where he received a BA in theology. His father had procured a position for him as an Anglican pastor, but Charles never was ordained and did not practice. More interested in natural history, he studied botany and geology and aspired to travel for study in the tropics, a popular avocation of young men of independent means. One of his professors, John S. Henslow, got him an unpaid position on the HMS Beagle as gentleman's companion to Captain Robert FitzRoy on a voyage to map the coastline of South America. Darwin spent his time collecting fossils, plants and animals from South America to the Galapagos Islands to Polynesia. Because Professor Henslow popularized the collections he sent back before his return, Darwin was a celebrity when he arrived home.

When Charles returned from his five-year around-the-world trip in 1836, he published detailed journals of the trip, as well as other scientific books, and delivered papers to the Geological, Geographical, and Zoological Societies. He spent another twenty years studying barnacles, pigeon breeding, and similar subjects. He never addressed evolution in any of these publications. He did not publish *On The Origin of Species* for twenty-three years! Supposedly he did not publish earlier because he feared reprisals, but being of independent means, being recognized as an authority in his field, and actively dialoging with leaders of the day about other theories of transmutation of species, this seems to be a thin excuse invented by later authors. This excuse was never alluded to in his book. Instead, he described working on it steadily over all those years and that he chose to publish his "abstract" (*On the Origin of Species*) due to failing health, although he said it would take three or four more years to complete his work.

It wasn't until Alfred Russell Wallace, a naturalist and admirer, sent Darwin his observations and theory of evolution while still away on a voyage to the Malay Archipelago and Borneo, that Darwin's theory was (hurriedly?) presented and published, establishing primacy over Wallace. To his credit, when his friend Charles Lyell presented the joint papers[19] to the Linnaean Society, Darwin acknowledged Wallace as co-founder of the theory. Claiming to have sat on his theory for over twenty years, he rushed to publish *On the Origin of Species,* which he described as an unfinished manuscript without supporting facts, acknowledgements or references.[20] For more details, see Appendix A for a short analysis and critique of *On the Origin of Species*, first edition, and a comparison to the sixth (and last) edition, which was only minimally changed from the first edition except for historical recognition of others before him and attempts to address some of the most important criticisms.

SHORT LIST OF EVOLUTIONARY THEORIES BEFORE DARWIN

WHEN	WHO	WHAT	HOW	DOCUMENTS
1726	Charles-Louis de Secondat, Baron de Montesquieu	Evolution	?	
1745	Pierre-Louis Moreau de Maupertuis	Evolution	natural selection of the fit	*Venus Physique*
1754	Denis Diderot	Evolution	change through natural selection	*Thoughts on the Interpretation of Nature*; Lettre sur les aveugles ("Letter on the Blind") (1749)
1772	James Burnett, Lord Monboddo	Evolution	natural selection of the fit	
1779	David Hume (d. 1776)	Evolution	refuted arguments from design and miracles	*Dialogs Concerning Natural Religion*
1749-1778	Georges-Louis Leclerc, Compte de Buffon	Evolution	natural selection of the fit	*Histoire naturelle, générale et particulière*, 36 volumes

1712-1778	Jean-Jacques Rousseau	Evolution	ape compared to man	
1794	Erasmus Darwin (Charles Darwin's grandfather)	Evolution	acquired characteristics	*Zoonomia*, 1794; *Temple of Nature*, 1803
1800	Jean-Baptiste Lamarck	Evolution	acquired characteristics	*Hydrogeologie*, 1801;
1813	William Charles Wells	Evolution	natural selection	"Account of a female of the white race... causes of the differences in color and form between... races of men" (Royal Society paper)
1818	Geoffery Saint-Hilaire	Evolution	unity of plan; environment causes changes	*Philosophie anatomique*
c.a. 1820	Robert Grant	Evolution	spontaneous changes and environment	
1831	Patrick Matthew	Evolution	natural selection	*Naval Timber and Arborculture*
1844	Robert Chambers	Evolution	acquired characteristics	*Vestiges of Creation*

I have always wondered whether Wallace was the true originator of a theory that Darwin had overlooked in his own observations, although he had written letters to Joseph Hooker and Asa Gray earlier hinting at an evolutionary theory. Did Wallace provide the link that brought all his speculations together? Because Darwin was backed up by his friends Joseph Hooker and Charles Lyell in his claim of primacy, we may never know. It is sure that the scientific reputation of Wallace declined, while Darwin's grew. It is interesting to note that Wallace later rejected the theory as lacking both mechanism and sufficient evidence. Others have also speculated about Wallace being the true originator of the theory.[21]

CHAPTER 9

EVOLUTION: THE THEORY, THE CLAIMS, THE EVIDENCE

Evolution, in its many grammatical forms, is one of the most overused, misused, and abused words in the English language today. It seems everything from ideas to stars to society is evolving today, which is in line with Progressive philosophy. Just what do we mean by evolution? Well, it has quite a few meanings. In a general sense, it is simply change over time, ie, history, with no reference to any scientific mechanism. It can also refer to any scientific, technological, political, or social system, philosophy, or business that develops, improves or changes throughout its history. For example, Wells Fargo has evolved from a gold rush banking, freight, stage coach, and security service to a modern international financial institution.

To true believers in Darwin's theory of evolution, it means that all creatures have descended from one or a few single-celled ancestors by random, gradual genetic modification over long periods of time, and that only the fittest survive through "natural selection," made up of factors such as competition, predation, disasters, isolation, and environmental changes. In the words of the theory, it is descent with modification through natural selection. Here for clarity, I will use the proper noun "Evolution" for the Darwinian type and "evolution" for everything else, though sparingly. I will also substitute other words such as development and change to avoid confusing redundancies.

The claim of the theory of Evolution is not just variability of offspring within a species as seen in breeding practices, from dogs to flowers to food crops, but the creation of totally new unique species (and advancing up the evolutionary tree from species and genus to phylum and kingdom) as far removed from each other as the whale and the hippopotamus.[1] Some refer to within-species variation as microevolution, and Darwinian Evolution, by contrast, as macroevolution. In arguments supporting Evolution, proponents often equivocate; that is, they interchange definitions in the middle of an argument to win by illegitimately adding an air of legitimacy. In particular, micro- and macroevolution are often used interchangeably as Evolution although they are two different phenomena.

If I have erred in giving to natural selection great power, which I am far from admitting, or having exaggerated its power, which is in itself probable, I have at least, as I hope, done good service in aiding to overthrow the dogma of separate creations.

Charles Darwin, Descent of Man

We know that Evolution's assumed means is defined as common descent with modification through natural selection, but is it science? **Our aim here is not to argue the merits of one theory over another, but to decide whether it is merely a scientifically based belief or science in pursuit of the truth.** Remember, science is the pursuit of truth about the predictable, repeatable and measurable aspects of the universe with which we can or could conceivably interact. To be science, it must be testable, verifiable and falsifiable. It should also make predictions about future outcomes.

Even if it is true that today's animals are descended from extinct animals of the past, is Darwinian Evolution science or philosophy? Is it founded on experimental and observational evidence for and against, or is it a belief system based on faith? What are the mechanisms? Is the evidence for them valid and sufficient? Is there reason to believe that the conclusions drawn from experimentation are plausible?

The fact that there is no other viable theory that excludes a creator as a possibility doesn't make it true or even science.[2] A Theory is generally an educated guess that fits observation better than any other explanation. In historical sciences,[3] it is called inference to the best explanation. But, is it plausible and can it pass the tests outlined above?

I have concluded that Evolution is not fully science because it relies too heavily on inference of conclusions that are not well connected or supported by the facts or experiments. Proponents argue that there is sufficient evidence to prove that it is true. Some others have said that the statement "natural selection through survival of the fittest" is merely a tautology or truism because the conclusion is self-evident within the statement. By definition, those that survive are the most fit to survive or they wouldn't survive. How is that anything but a statement with circular reasoning?

The theory of Evolution relies heavily on rationalism rather than empiricism, on deductive reasoning rather than induction from experiment or observation. Empiricism is all about experimentation and facts. Rationalism is all about drawing conclusions about what must have happened, based on mental or logical principles. In science, rationalism must always be supported by empirical results. Since we cannot go back in time to recreate or test events, Evolution is necessarily more about conclusions drawn from forensics than about facts revealed through experimentation.

What about descent with modification? Is there direct evidence of that? The only directly connected evidence of descent with modification is in modern experimental genetics and breeding. Both of these have failed to produce a single new species, much less a new genus, despite more than a hundred years of breeding experiments on short life span species (eg, fruit flies, plants, yeasts, bacteria). They have only produced varieties of a given species. Breeders, including Darwin, have long known that the spectrum of such variation is limited about a norm and often returns to original type.

For example, breeding fruit flies may result in red or black eyes, short or long wings, but not another fly species. Species, by definition, are sterile with respect to natural breeding with other species. Hybridization is the crossing of distinctly different species. Hybrids are generally sterile, and new crossings of the original species must be made to produce viable seeds or offspring. Many of our modern commercial crops are hybrids so that new hybrid seed must be purchased each year. Crossing a horse and a donkey produces a sterile mule, not a fertile half-horse, half-donkey. To produce more mules, more crossings are necessary.

One important source of indirect evidence is the fossil record, which is anything but a continuous series of small modifications from one species to another. In contrast, new species seem to arise without precedent, presumably due to the incompleteness of the fossil record. Darwin knew of this discontinuous fossil record, but assumed that the gaps would eventually be filled by more exploration. This never happened.

> I have read your book with more pain than pleasure. Parts of it I admired greatly; parts I laughed at till my sides were almost sore; other parts I read with absolute sorrow; because I think them utterly false & grievously mischievous — You have deserted—after a start in that tram-road of all solid physical truth—the true method of induction—& started up a machinery as wild I think as Bishop Wilkin's locomotive that was to sail with us to the Moon. Many of your wide conclusions are based upon assumptions which can neither be proved nor disproved. Why then express them in the language & arrangements of philosophical induction?"
>
> *- Adam Sedgewick, noted geologist who had taught Darwin, after reading Origin of Species*

The fossil record shows that animals existed in the past that were different from those that live today, but there is not one, even remotely, complete series showing gradual change into another species. Despite the attempts to connect fossils to living species, such as the famous horse or man lineages, none have really been connected by discrete small changes from one to another. Similarity of form and large leaps are said to prove small changes and direct connections. It is all inference from incomplete data to fit the preferred theory. This is not to say that there have not been connections, but that they are far from being proven.

Instead, Darwinists have always relied on similarity of form (homology) to imply descent from one to the other. It is interesting to note that there are countless species living today that have remained relatively unchanged for most of Earth's history. Yet this is not seen as a falsification of the theory. They are merely assumed to have been extraordinarily fit for all of the changes the earth has seen. These panchronic[4] species include such delicate creatures as frogs and algae and are supposed to have survived ice ages, meteorite strikes and volcanoes. Yet, today we are told that pollution or a couple of degrees shift in temperature will cause their extinction. What's wrong with this picture? **Panchronic species include some members of the following groups (short list):** Dragonfly, cockroach, ant, termite, fruit fly, mosquito, beetle, flea, spider, scorpion, centipede, crocodile, frog, snake, shark, jellyfish, coral, squid, starfish, horseshoe crab, bony fish, lung fish, fern, ginkgo, mushroom, bacteria, yeast, mold, single cell and colony organisms, algae, and lichens.

Recently, biochemists have compared DNA gene sequences and the proteins for which they encode. These do seem to track with homology in that similar animals living in similar circumstances have similar genes and enzymatic proteins. To imply a date of separation from a common lineage, they estimate the (assumed) rate of mutation and count the differences in the molecules. Unfortunately, these projected dates do not agree within the same organism for different gene markers. Such estimates rely, at least in part, on the age of assumed fossilized common ancestors to set the assumed mutation rates. This is circular reasoning. The link is assumed and then the differences are fitted to it afterwards.

Only about 1.5 percent of genes encode for proteins that make up the thousands of enzymes and cell structures. Most of the rest of the DNA is poorly or not understood at this time.[5] Because internal metabolic functions at the cellular level must necessarily be similar in all species for life to exist, and enzyme systems involved in these processes are so structurally specific to their functions, it is not surprising that they are so similar; it is surprising that there are any differences at all in these DNA sequences and the proteins (enzymes) they encode.

It was easy to believe in Evolution when our knowledge of the world, the nature of cells and the fossil record was sketchy. When Charles Darwin first proposed his theory, cells were thought to be simple bags of gelatin. No one knew of the complex molecules like DNA, RNA, and metabolic enzymes, or structures such as mitochondria, Golgi apparatus or genes that carry on a galaxy of mind-bogglingly complex processes inside each cell.

Darwin believed that hereditary information was passed down through "gemmules" that each cell in the body shed and that then migrated to the reproductive cells. This belief is called *pangenesis* because every cell in the body got a vote in the outcome. By this scheme, inheritance of acquired characteristics seemed plausible. Darwin emphasized use and disuse as causing Evolutionary changes. Advances in the knowledge of the biology and biochemistry of the cell make it harder and harder to believe in Evolution as it had originally been presented. In recognition of this uncomfortable situation, in the twentieth century a revision of the theory was agreed upon among evolutionists. It is known as Neo-Darwinism, and included biochemical similarities and concepts of molecular evolution.

> I am actually weary of telling people that I do not pretend to adduce direct evidence of one species changing into another, but I **believe** that this view is in the main correct, because so many phenomena can thus be grouped and explained.
>
> *- Darwin, letters (emphasis added)*

Could Evolution be true? Maybe. Is it fact or belief? At this point, it is a belief. Only the atheist NEEDS it to be true to "prove" that a creator is not needed. However, those who believe in a higher power don't NEED it to be false, because a being that could create a universe and life itself could surely use this means to create new species. Is it science? No, it is a scientifically based belief system, a philosophy. What would be needed for it to become science?

To be scientific, it must deal only with the matter or processes with which we can or possibly could interact, it must be verifiable, falsifiable and testable, and it should predict future outcomes based on experimental data. What question could possibly be asked to make Evolution falsifiable? Until recently we would say none. As we have seen, Evolution allows for survival of panchronic species throughout time, but also allows for change through slight modifications. Neither can falsify the philosophy, so it is not science.

But, wait a minute. What about the biochemical complexity? If you include that, and allow that some structures or processes could not possibly be built up through generations of gradual change, could that falsify Evolution? Potentially, yes.

If it could be demonstrated that any complex organ existed which could not possibly have been formed by numerous, successive, slight modifications, my theory would absolutely break down. But I can find out no such case.

- Charles Darwin

If it could be proved that any part of the structure of any one species had been formed for the exclusive good of another species, it would annihilate my theory, for such could not have been produced through natural selection.

- Charles Darwin

The almost infinitely complex biochemistry of the cell, rather than supporting Evolution, is the greatest challenge to it. The more that we discover about the biochemistry of processes within living things, the less likely it is that undirected Evolution is the only or any explanation at all. Since each gradual step in a series of changes would have to make the organism more fit than its predecessors many generations before the resulting structure fills the function it will ultimately perform, any step that is not advantageous would not be selected, according to the theory.

OK, then if each step would exist well before the ultimate function, wouldn't there be in each organism on earth a lot of junk DNA or nonfunctioning proteins or organs? The answer seems obvious: yes. Do we find this? No. As for the proteins that perform all of the functions of the cell, useless or interfering material is usually only found in defective cells causing disease, eg, PKU or Sickle Cell Anemia. Useless enzymes are quickly destroyed if they are formed at all. Enzymes are only produced in response to current needs as defined by a complex system of feedback communication within the cell.

As for DNA, we don't know yet whether there are actual proto-genes lurking there. Genes, which are sections of DNA wrapped around companion proteins (Histones), are separated by sections of so-called junk DNA (introns) for which we, as yet, know next to nothing about their functions. Could introns be the missing nonfunctioning proto-genes? Possibly. And that is the assumption of some Evolutionists. But remember that any junk must also have a survival advantage over predecessors, so it must function in some way. Regardless of what you may have read in the popular press, we are just scratching the surface on understanding the DNA code (or codes).

The only sections we really know anything about are those that encode for proteins used as either structural elements or enzymes that perform various cell functions, along with certain signals such as start, stop, zip and unzip used in replication or repair. We don't even understand why the units encoding for certain enzymes are broken up into sections, sometimes on separate genes along the DNA strand or why the RNA that is assembled from the DNA template is usually cut and/or rearranged before it carries the instructions from the nucleus to the ribosomes where the proteins are assembled from component amino acids.

Development is another important area where we lack understanding. How does a fertilized egg differentiate to form a complete creature? We can describe what happens at each step, but have yet to formulate a comprehensive theory of how the instructions are timed, given, and received. Much research is being done in this area, but results are sketchy and incomplete at this time.

INTELLIGENT DESIGN AND CREATIONISM (S)

If Evolution is a philosophy or belief system and not fully science, what about Intelligent Design, and what is it anyway? Intelligent Design (ID), like Evolution, is a philosophy that uses science as its basis. It is a theory that the very complexity and improbability of the universe and life processes necessarily implies an intelligence or greater principle behind it all. ID does not say what (or who) that intelligence is or connect it with any particular faith based belief system. ID proponents accept the possibility that some of Evolution's claims are true, but challenge the validity of other conclusions, particularly unguided chance as the only source of biological progress.

If gradual assembly, step by step, of complex systems, from DNA to hearts, by random chance is statistically impossible or implausible, it rules out chance as a cause. That leaves necessity and design as possible causes. ID contends that the very implausibility and complexity strongly imply that some intelligence (design) or as yet undiscovered guiding principle (necessity) must be at work to overcome the statistical barriers. Necessity fits in with the Progressive philosophy of inevitable progress built into the universe. Neo-Darwinian Evolution states that life and the universe only LOOK designed.[6] ID is the most significant challenger to Evolutionary theory at this time and some of the observations and experimentally derived facts could eventually lead to falsification or at least modification of Evolution as it is defined today.

The media tend to lump ID and Creationism together because the Evolutionists associate them in an attempt to discredit ID. ID accepts all scientific facts, whether it is the age of the universe and the ages assigned to the fossil record or biochemical or astronomical knowledge. ID acknowledges the appearance of design and infers that it necessarily implies an intelligence or guiding principle. The implication is that of purpose, not blind purposeless chance. This teleological[7] view is one of the things that makes ID a philosophy, not a hard science. However, most of the proponents are competent scientists who use valid scientific principles in scientific studies into the validity or lack of validity of Darwinian Evolution's random chance claims. Whether Darwinian Evolution or ID is more valid, is a matter of opinion at this point. Expressed differently, ID could be an attempt to falsify Evolution as arising only from random chance. Statistical arguments of the improbability of random processes explaining life and complexity are quite convincing. Note that Progressivism, espoused by Darwinians, also implies teleology because everything is believed to be naturally progressing toward perfection.

Biblical Creationism comes in two major forms: Young Earth and Old Earth. Young Earth Creationism declares that the universe and the Earth were created some 6,000 years ago in seven twenty-four-hour days by a strictly literal interpretation of the Biblical Genesis. Old Earth Creationism accepts the ages of the universe, the Earth and the fossil record and interprets the Biblical Genesis account somewhat more figuratively than literally. As such, it is closer to Intelligent Design, but specifies the God of the Bible as the intelligence responsible for it all. A subdivision within Old Earth Creationism is one that assumes a literal seven-day creation (or seven epochs) after the initial billions of years needed to form the universe and prepare the earth for that event. "In the beginning, God created the heavens and the earth," Genesis 1:1, is assumed to contain all of the history of the universe, the Earth and the fossil record that had already taken place before the creation event that formed modern humans. This is sometimes called the Gap Theory, especially by its opponents. I tend to fall into the Old Earth camp, based on scientific probabilities and evidence, while leaving the length of the seven days as an open question. The same word is used for a twenty-four-hour period and for an indefinite time period as "in the day of Moses." "These are the Generations of the heavens and of the earth when they were created, in the day that the Lord God made the earth and the heavens." Genesis 2:4. Also, in 2 Peter 3:8[8] it is obvious that they did not limit a day to twenty-four hours.

It is easy to see that Creationism, based not on science but biblical texts, is different from both Evolution and Intelligent Design. While it is true that some who believe in Intelligent Design are also Old Earth Creationists, it is not true for most of ID's technologically and scientifically astute proponents. As a matter of fact, many proponents of Creationism, especially Young Earth Creationists, oppose Intelligent Design as a distortion and means of explaining away the literal interpretation of biblical texts that they hold to be infallible. Questioning the literal interpretation of Genesis is seen by them as opposing all of God's truth in His Word. In their opinion, rejection of literal seven-day creation ex nihilo (from nothing) is tantamount to saying the biblical texts, in their entirety, cannot be trusted to be true.

CONCLUSIONS:

So, where does that leave us? We have three different philosophies for the development of the complexities of life, at least two of which base their beliefs on scientific inquiry and knowledge. Which is correct? The jury is still out. Neo-Darwinists see Intelligent Design, with its underpinnings of biochemical complexity and probability theory, as the greatest threat to their beliefs and rightly so. Rather that embrace Evolutionary philosophy and all its claims, ID proponents critically examine the evidence for and against guided and unguided development, including the incredibly complex biochemical nature of life and statistical probabilities of unguided or chance development of complex molecules, structures and functions. If evolution is a robust theory and the aim is truth, such a challenge should be welcomed as a further development of understanding Evolution. The vitriol with which Evolutionists attack ID proponents can only be interpreted as religiously defending dogma that they feel may be vulnerable.

PART 4

ORIGIN OF LIFE THEORIES AS MOLECULAR EVOLUTION

CHAPTER 10

LIFE ITSELF

ife. What is it? We can describe the characteristics and infinitely varied forms of living things, but what exactly is life itself? In the past, it was assumed that there was a vital force present in all living things that passed down from life to life. This philosophy was called *vitalism.* Because it borders on the divine, today vitalism has been replaced by the philosophy of *mechanism*, which states that all natural phenomena, particularly life, can ultimately be explained through physics and chemistry. The universe is thought to be merely mechanical in nature. Under this philosophy, life is just a process produced by physical laws acting on matter.

Life is assumed to be a given, like gravity, which incidentally has not been explained or understood well either. We know gravity exists and how it behaves, but really don't understand why. There is not a single location in any creature or cell that we can point to and say, this is its life. Definitions of life usually describe what living things consist of and what they do; they do not actually tell us what life itself is. We can't collect, isolate, or test it, so it appears to be a transcendent quality. What exactly distinguishes a living cell from a dead one or a mixture of cellular components? Depending on the source, explanations vary from biochemical to functional.

Life only comes from life. Spontaneous generation of living things has been shown over and over to be false. Spoiled food does not beget flies or mold. Each only comes from other flies or mold spores. Life as a process requires just the right kind and amount of regulated energy and a fine balance of the right molecules and structures. Science has not been able to create life or even most biological molecules without the help of molecules first derived from living systems or those systems themselves, eg, bacteria engineered to produce insulin. Even if all of the components of a living organism are blended in the lab in the correct proportions, no life results.

What is it that assembles and winds up the machine or provides the vital spark? Science does not know. Proponents of molecular evolution believe that non-living molecules at least once in Earth's history spontaneously became a living system from which all subsequent life descended. They argue (and with some merit) that spontaneous generation cannot occur today because living organisms would consume any components before they would have time to accumulate and self-organize into a living system. They assume that only in the absence of life could components accumulate sufficiently to create life spontaneously from non-living components.

Never mind that the key molecules, eg, proteins and nucleic acids, are unstable in water for the length of time that would be necessary to accumulate and assemble the correct mixture into a living system. These molecules are assembled by linking smaller molecules together with the loss of a water molecule for each link. When excess water is introduced, eg, ocean, the reactions tend to be reversed and the links fall apart. That is why proteins inside cells are constantly being assembled to replace those that have been degraded. Molecular evolution proponents believe that production of life in the laboratory can be accomplished at some future time, although they have no evidence to support that belief. We will look at some of the more popular origin of life theories and the validity of the arguments later.

Life is a continuous process that is constantly working against forces that would end it. It has been said that life 1) is improbable, 2) defies entropy (the Second Law of Thermodynamics), 3) is unstable, and 4) needs a constant supply of raw materials and energy to survive. Let's look at each of these claims.

1. LIFE IS IMPROBABLE

Life really is improbable, partly because of the extremely low probabilities of such complex systems forming by random chance even once. It is the ultimate "infinite improbability drive."[1] Even the simplest known bacterium contains thousands of types of proteins and other unique biological molecules and structures. Metabolic processes necessary for life depend on thousands of different, specific enzymes for facilitation and regulation through feedback and other means. Enzymes are proteins made of chains of amino acids that are folded into useful shapes. If we suppose that an average enzyme is 200 amino acids long,[2] using the 20 left-handed amino acids living beings use, the probability of only one specific enzyme sequence forming at random is 1 in 20^{200} or 1 in 10^{260}. That's a 1 with 260 zeros after it.

If the universe is 13.7 billion years old, there have been 4.32×10^{17} seconds since it began. We would need to make 231.4×10^{180} attempts each second since the beginning of the universe to make the random assembly of even this one specific protein plausible – that is, to make the number of attempts that are in the same ball park as the odds. This all pre-supposes that all of the amino acids have been pre-assembled and are readily available. Amino acids occur in left- and right-handed forms, and only left-handed forms are used by living things. If we take this into account, the odds would be much higher. But remember, to create even one cell all this must happen in a very confined space so that all of the proteins and other molecules can be collected in one place, not just anywhere in the universe or even anywhere on the surface of the Earth.

If we assume that life molecules were assembled on Earth, which is thought to be only 4.5 billion years old, and evidence of life was present 3.8 billion years ago, then the number of attempts per second rises to even more impossible levels. And that is just for one protein enzyme assembled from readily available units, excluding interfering molecules, and under the ideal conditions for assembly and preservation. Already we are seeing the extreme odds against a specific enzyme being produced. If we look at what it would take to produce by chance the thousands of different specific enzymes necessary for metabolism, the probability of random assembly of the correct mix would be $(20^{200})^{3000}$ for a simple bacterium with 3,000 enzymes, or 1 in $10^{780,000}$; that's a 1 with 780,000 zeros after it. The terms *impossible* and *miracle* come to mind.

Now let us look at DNA. There are four different molecules that create base pairs like the rungs on a ladder along the coiled double helix of DNA that encodes for proteins. Bacterial DNA, whose chain forms a circle and is tightly wound around proteins, is 300,000 to 4 million base pairs in length. If we assume that a simple bacterium has DNA that is 500,000 nucleotides long, using 4 types of bases (two purines and two pyrimidines), the probability of forming the correct sequence is 1 in $4^{500,000}$ or 1 in $10^{301,030}$ – that's a 1 with 301,030 zeros after it. Even this presumes that each nucleotide has already been pre-formed from one of the four readily available bases, its partner and a pair of specific phosphorylated sugar (deoxyribose) molecules that create the sides of the ladder.

It's even worse than that, however, since each purine must pair with its specific pyrimidine to create each Base pair[3] so double the number is needed. Now add the probabilities of assembling, in one place, the DNA and its associated proteins (histones), the thousands of enzymes and other structures like cell membranes, and it is obvious that the probability of forming even the simplest bacterium is so infinitesimally small that it can only be called either impossible or a miracle. Even if we assume that an earlier form contained a tenth or a hundredth of this number of components, it would still be called impossible or a miracle. For 1 percent of the components, it would be 1 in $(10^{260})^{30}$ or 1 in 10^{7800} (1 with 7800 zeros) for enzymes and 1 in 10^{3010} (1 with 3,010 zeros) for DNA (or RNA), plus assembly of all the other components as noted above.

One of the origin of life theories proposes that RNA, not DNA was the original control and inheritance molecule. The difference in the structures of DNA and RNA is that DNA uses the deoxy- form of ribose and RNA uses ribose itself. Since DNA now transcribes instructions for protein assembly to RNA first, the theory skips this extra complexity as a more believable scenario. Presently, some viruses use RNA instead of DNA, but viruses are incapable of most life processes on their own and must take over the DNA of host cells to reproduce. They can be thought of as parasitic seeds, not complete organisms.

Fred Hoyle, a famous astronomer and atheist, stated that the odds of forming a living being at random from lifeless molecules would be like the chance that "a tornado sweeping through a junk-yard might assemble a Boeing 747 from the materials therein."[4] Note that Fred Hoyle and N. C. Wickramasinghe[5] estimated the odds at 1 in $10^{40,000}$ by assuming that numerous structures of enzymes could perform the same functions. That's still pretty steep odds. Others have calculated the odds with various assumptions and outcomes, but all result in extremely small odds.

All of the extreme improbabilities don't even address whether life would spontaneously arise under the right conditions, if all components are available, or whether we would just have the same non-living jumble of molecules we could assemble in a laboratory. In other words, we still haven't addressed what assembles and winds up the mechanism to start life processes. Clearly, some other unknown process besides random chance has been at work in both assembling the components and in turning them into something alive.

2. LIFE DEFIES ENTROPY
(SECOND LAW OF THERMODYNAMICS)

Entropy is a measure of the disorder of a closed system and the Second Law of Thermodynamics states that entropy always increases – that chaos always increases and usable energy always decreases. Life seems to defy entropy because life is very, very organized and uses matter to generate energy and build more and more complex structures. However, living things are never closed systems; they need material and/or energy from outside to survive, so an organism that seems to decrease entropy within itself may increase entropy of its surroundings continually. Is it enough to result in a net increase in entropy of the earth or the universe? The answer is unknown but possible. Note that this assumes that the Second Law of Thermodynamics is absolutely true in all cases, but this has not been proven either. It is a well-accepted and thoroughly tested theory and thus is a scientific law by that definition.

A planet with abundant life is far more complex and organized than a dead planet simply because the chemistry of life is far more complex and dynamic than inorganic crystalline structures. It is difficult to see how the net decrease in entropy caused by life on an isolated planet can affect any other planet, much less the universe as a whole. If the planet is considered the "closed system" then there is indeed a net decrease in entropy and a net increase in complexity, order and usable energy, eg, fossil fuels. Of course, that also depends on your definition of order and disorder. If we define disorder as an increase in the number of states, and order as uniformity of form and function, than the dead planet is not as disordered as a planet with abundant life in all its forms and complexity of functions. However, if disorder is the rule, then the ultimate outcome of continued disordering and loss of energy is a uniform, cold, dead universe in the lowest energy and organization state possible.

3. LIFE IS UNSTABLE

Life is indeed unstable because it exists on the edge of destruction, far from equilibrium. Ordinarily, chemical reactions reach a state of lowest energy called *equilibrium* where they are stable. At that point the reaction stops or is stabilized dynamically where the net amount of products no longer increases and the net amount of starting materials no longer decreases. Life is never at or near equilibrium and requires input of material and energy to maintain itself in this unstable state. It can only exist under very specific physical circumstances including temperature, pH, pressure and presence or absence of oxygen. An aerobic organism requires oxygen, whereas oxygen is deadly to an anaerobic organism. The only time an organism is stable or at equilibrium is when it is dead.

4. LIFE NEEDS A CONSTANT SUPPLY OF RAW MATERIALS AND ENERGY TO SURVIVE

Life requires a constant or near constant supply of materials and energy from outside itself to survive. Ultimately, most of life on Earth depends on the products of photosynthesis as a source of energy that is initially derived from the sun. The only exceptions are those living systems present in deep seas and deep interiors that derive energy from bacterial processing of inorganic chemicals such as hydrogen sulfide. In both cases, energy and material from outside the organism are necessary to maintain life.

Since no one knows what life actually is, the best we can do is define what living things must have and must do to live. All living things are more alike than different. An advertising flyer I received a few years ago from a supplier of products for biochemistry stated, "Did you know that humans share about 50 percent of their DNA with bananas?"[6] All living things use essentially the same basic biochemical processes such as metabolism in the everyday business of living, so the DNA that encodes for the chemicals used for life processes are necessarily very similar. The differences are relatively minor compared to the similarities. The processes used to accomplish all of life's functions at the molecular or cellular level have to be very similar for all living beings. Because the processes are so complex and similar, the surprising thing is not that the workhorse protein molecules (and thus the DNA that encodes for them) of different living things are so similar, but that they are as different as they are and still function in essentially the same way.

Living things at the minimum consume and process food, excrete waste, grow and reproduce. Some evolutionists would add "and, through natural selection, adapt in succeeding generations."[7] Some living things also move, sense and communicate. Some can even go dormant for long periods and only "come to life" when conditions are right. This is true of many bacteria. Bacteria that had lain dormant for 120,000 years have been found under Greenland's glaciers.[8] I once left a closed jar of saturated salt solution, which I had used to treat a sore throat, sitting for a month or so. When I started to throw it out, there was a fuzzy white ball of bacteria floating in the middle of it. This extremophile[9] bacterium that could grow in this high salinity environment was probably from the salt and may have been dormant for thousands of years before awakening.[10] Revitalization of dormant organisms is a great mystery. How can life itself be suspended and then be restarted spontaneously? Is it really suspended, or is it just slowed to an imperceptible level? But how could it survive for thousands of years?

CHAPTER 11

BIOCHEMICAL COMPLEXITY
OF LIVING THINGS

All living things consist of one or more highly organized cellular units containing nucleic acids, proteins, lipids (oils), sugars, and a complex mixture of other carbon-based organic chemicals, in a water base, separated from the environment by a cell membrane that regulates what passes in and out. Cells use materials and/or energy from the environment to perform chemical or physical work and/or build other molecules and excrete waste products. Cells must contain:

Proteins for structure and for various functions such as metabolism;

Nucleic Acids for replication, for regulation of energy and processes, and for instructions to build proteins;

Sugars for energy storage and generation and for building other structures such as starches and cell walls;

Lipids (fatty molecules) for energy storage and for membranes.

Phosphate ions attached to other molecules to store and release energy in a very controlled way;

Internal structures for metabolism, respiration and other functions necessary for life.

Biochemical processes inside every cell of every creature are the most complex and convoluted processes known. Inside every cell of your body, thousands of biochemical processes are taking place every second. Molecules are constantly being built, used, recycled and destroyed. Special protein molecules called enzymes facilitate most steps of these complex, multi-step processes. These in turn are produced by instructions from the nucleic acids DNA and RNA and are transported in the right amount needed to the exact site where they are used.

Inside cells there are countless examples of systems and processes with complex interactions that are so interdependent that it is difficult to imagine how they could have developed piecemeal over time. It's like the chicken and the egg dilemma. Without the chicken you can't have eggs, but without an egg you can't have a chicken. So, which came first? Complex, interactive systems are like that simplified example. Without DNA you can't have proteins, but without proteins, you can't produce or maintain DNA. Which came first? This question has baffled scientists for generations. Some say a protein-based system developed first and evolved into a system based on nucleic acids. Others favor a nucleic acid system that evolved the ability to produce and develop proteins, known as the RNA world hypothesis. All of this is, in fact, speculation with little or no evidence or basis in fact. Basically, it is a great mystery.

The biochemistry and structure of a single, basic cell is a wonder that is so complicated that most processes could not function if even one of the hundreds of parts were missing, incomplete or defective. For even a simple bacterium to live, it must use food to produce energy and build or regenerate structures. Let's look at one of the most simple and basic processes, turning a simple sugar into energy. What follows is a simplified summary of an extremely complex process.

If you want to know more of the details, find a good introductory biochemistry book in your local library or college bookstore. There are whole chapters on the subject.

GETTING ENERGY FROM SUGAR

For the cell to burn a simple sugar (glucose) for energy, it takes nineteen steps in three phases. Each step is controlled by a different enzyme.

Enzymes are huge, complex protein molecules that speed up or slow down the chemical reactions taking place. Think of them as huge machines that hold, cut, combine or rearrange other molecules. In this example, when available energy is lowest, two of the nine enzymes in the first phase (glycolysis) accelerate the process, but they also slow or shut it down when energy reserves are high. This way, sugar is only used up when it is needed for energy production.

The first phase turns one sugar molecule (glucose, having six carbons) into two smaller molecules (pyruvate, having three carbons). It uses two energy units but produces four. ATP (adenosine triphosphate) is the high energy unit in this phase. Each time ATP gives up energy to another process, it loses one phosphate ion, becoming ADP (adenosine diphosphate) and then loses another to become AMP (adenosine monophosphate).

More ATP is produced by adding back phosphate ions to AMP or ADP.

The second phase uses a coenzyme[1] to cut the result of phase one (pyruvate) into an even smaller unit (acetyl, having two carbons) and carries it to phase three.

The third phase is called the citric acid cycle because it cycles through all of the steps but starts by adding the acetyl ion to another molecule (oxaloacetate having four carbons) to make citrate (a form of citric acid having six carbons). When citrate is high, it regulates the process by slowing or stopping the first phase, which in turn feeds the other two phases. In phase three it takes nine steps with eight enzymes that require six different cofactors (other molecules that turn the enzymes on or off) to produce carbon dioxide and one energy unit[2] for each acetyl unit. Only the first step of this phase is not reversible. The other steps can be slowed, stopped or reversed by an abundance of any of their products, or an abundance of the high energy units.

Most if not all biochemical processes involve feedback loops that speed up, slow, stop, or reverse steps in the process. Without this feedback, the result could be too much or too little energy or other desired products. If any of the enzymes, energy units or building blocks is missing or damaged, this process simply won't happen and the cell will die from lack of energy or a build-up of harmful by-products. All of this is happening thousands of times a second in every cell of your body. See Appendix B for a more detailed picture of this process and the enzymes and other molecules involved.

If you looked at the basic chemistry without considering the regulatory enzymes that facilitate and regulated it, you might say it could happen by chance. However, when the interdependence and complexity of the many enzymes and the feedback loops are considered, it is impossible to imagine this system developing in any way but all at once. Enzymes often are stored for short times as proenzymes, which are inactive forms that must be activated by some other agent just before they are needed. Enzymes are constantly falling apart or being dismantled by still other molecules so they must be produced constantly by the cell to be available for use.

Could you break down sugar into carbon dioxide and energy without the enzymes? Yes. Then why are the enzymes necessary? There are three major reasons. First of all, without enzymes many steps would either take too long (weeks or years) to be practical or they would need too much energy (heat) to be compatible with living cells. Biological processes using enzymes are like boiling an egg in your hand instantly with only the heat from your body and room temperature air. Without some special agent, it simply wouldn't happen in our lifetime. The process is simply too slow or impossible without a considerable amount of heat being added. The second reason is the regulation by feedback that produces just the right amount of each product and energy at just the right time.

The third reason is that the energy units (nucleotides like ATP and GTP) store and carry energy for use in just the right step of each precisely controlled process. In a typical chemical reaction, any energy produced would exist as radiation such as heat or light, so it either wouldn't stick around long enough to do anything else, or because it would not be controlled or directed, it would do even more damage than the heat used to make the reaction go in the first place. Since enzymes are made of proteins and proteins are easily damaged by heat, excess heat would just cook them (like the egg, whose white is mostly protein in water).

If the basic process of turning sugar into carbon dioxide and energy is this complex, then you would expect other processes like building new enzymes or DNA or whole cells must be far more complex. As a matter of fact, they are. I'll give just a couple of classic examples here.

Example 1 – Photosynthesis has been present on earth from its earliest times. Some of the oldest known fossils are of photosynthetic cyanobacteria, formerly known as blue-green algae. Although the process of photosynthesis is somewhat simpler in these simple organisms than in larger plants, each process is very similar and complex. Basically, photosynthesis harvests sunlight to synthesize sugars from carbon dioxide and water using an antenna system to collect the sun's energy and a cascade of reactions to turn carbon dioxide and water into sugars. These sugars are also the starting materials used to build more complex molecules including fats, starch and cellulose. I have not included the details here for brevity, but if you are interested, you can find the amazing details in a basic Biology or Biochemistry textbook.

Example 2 – Building a protein from amino acids starts with sections of DNA being transcribed into RNA. Sometimes several separate sections of DNA are used to assemble the RNA. The RNA thus assembled is often cut, shortened, resectioned, rearranged or combined with other molecules by enzymes into its final form, called mRNA (messenger RNA) before being transported from the nucleus (DNA area or nucleoid in bacteria, which has no nucleus) to a Ribosome where assembly of protein takes place. A Ribosome consists of proteins and two other types of RNA called tRNA and rRNA (transfer RNA and ribosomal RNA) that transfer amino acids or catalyze assembly of the growing protein chain.

The final form of the mRNA is then read by the Ribosome, which adds one amino acid molecule for each set of three base pairs from the RNA to the growing protein chain. Each amino acid is coded by three unique base pairs in the RNA. Once the protein is assembled, it may be further modified, folded or rearranged by other enzymes to form the final enzyme or structural component needed. Then it is transported to the area where it is needed to perform the work it is designed to do. Afterwards, when it is no longer needed or is actually detrimental to other processes, it begins to deteriorate and/or is disassembled. The amino acids are often reused to assemble other proteins.

Behind all of this are molecules and mechanisms that:

1. "Tell" the DNA that the protein is needed and to begin the process,
2. Locate the right sections of DNA,
3. Unzip the DNA to
4. Begin replication on one leg of the "ladder" into RNA,
5. Assemble, gather and transport the nucleotide bases needed to assemble the RNA,
6. Release the assembled RNA from the DNA,
7. Re-zip the DNA after replication,
8. Modify the RNA into its final form,
9. Transport the RNA to the ribosome,
10. Provide the energy for all of the operations,
11. Assemble and transport the correct amino acids to the ribosome to
12. Assemble the protein according to the RNA instructions,
13. Modify the protein to its final useful form,
14. Recognize the need to disassemble the RNA after use,
15. Dismantle or recycle the nucleotide base units and
16. Transport them to where they are needed,
17. Move the new protein to the site where it is needed,
18. Disassemble the protein when it is no longer needed.

COMPLEXITY OF CELLS, THE BASIC UNITS OF LIVING THINGS

The biochemical processes and the structures where they occur in each cell are so complex that the cell can be considered not just as a factory but an entire city with specialized businesses dedicated to functions such as manufacturing, storage, recycling, import, export and transport of raw materials and products, disposal of waste, construction and demolition. A listing of structures inside a typical cell where these processes occur includes the following necessarily shortened list and simplified definitions:

CELLULAR STRUCTURES AND THEIR FUNCTIONS:

1. **Plasma Membrane** – the (phospho)lipid bilayer enclosing the cell or other structures within to separate them from their surroundings.

2. **Cell Membrane** – the plasma membrane enclosing the cell having special protein receptors and portals for regulating passage of materials through it. It may also contain structures for mobility such as cilia and flagella, as well as other specialized structures such as those for recognition of foreign invaders that trigger immunity.

3. **Cell Wall** – a rigid structure enclosing the cell, including its cell membrane, usually made of complex sugars; uncommon in animals, but usually present in bacteria and plant cells.

4. **Plasmodermata** – Openings in plant cell walls that allow direct contact between cell membranes and cytosol.

5. **Cell Capsule** – a gelatinous structure of complex sugars, present in some bacteria enclosing the cell wall and the cell membrane, and that gives added protection.

6. **Mesosome** – in bacteria, plasma membrane folded in on itself for various functions.

7. **Capsid** – Protein coat of viruses enclosing DNA and/or RNA; used to inject its contents into a host cell for replication

8. **Intercellular junctions** – used to hold cells together in tissues; three types are tight (cell membranes fused), desmosome (anchored by keratin) and gap (protein regulated communication) junctions.

9. **Nucleus** – in eukaryotes, the central membrane-enclosed body containing DNA and associated reproductive and regulatory structures for replication, regulation and protein templates.

10. **Nucleolus** – a structure inside the nucleus that assembles ribosomes

11. **Nuclear Envelope** – plasma membrane enclosing the nucleus, studded with nuclear pores and connected to the endoplasmic reticulum

12. **Nuclear Pore** – protein enclosed opening in the nuclear envelope that regulates passage in and out.

13. **Nuclear Lamina** – netlike structure inside the nucleus for support and structure

14. **Nucleoid** – in bacteria, central concentration of DNA not separated by a nuclear envelope.

15. **Rough Endoplasmic Reticulum** – complex of folded membrane enclosed spaces, continuous with the nuclear envelope, which is

studded with ribosomes for protein synthesis. Also involved in transport of products and calcium up-take.

16. **Ribosome** – specialized protein machine containing short segments of RNA that synthesizes proteins following instructions from RNA replicated from DNA segments. Some are attached to the outer surface of the membrane enclosed spaces, but others reside in the cytosol (cell fluid).

17. **Smooth Endoplasmic Reticulum** – Complex of folded membrane enclosed spaces connected to the rough endoplasmic reticulum that synthesizes steroid hormones and lipids (fats and oils) in liposomes; detoxifies harmful chemicals; removes phosphate from phosphorylated glucose for transport and use; transports and packages proteins, fats, et cetera.

18. **Liposome** – specialized structure that synthesizes lipids such as fats and fatty acids.

19. **Golgi Apparatus** – Complex of folded membrane enclosed spaces unconnected to the endoplasmic reticulum that receives proteins and lipids in vesicles from the endoplasmic reticulum and packages them for transport in independent vesicles.

20. **Vesicle** – any membrane enclosed space used to store or transport products such as proteins, lipids, water or waste.

21. **Transport Vesicle** – small membrane enclosed space budded off from the Golgi complex that moves materials from the sites of synthesis and processing to the place where they are used.

22. **Secretory Vesicle** – like transport vesicle, but that moves products through the Plasma Membrane to the outside of the cell.

23. **Vacuole** – a membrane enclosed space where food, water or waste is stored until needed or until disposal.

24. **Lysosome** – special structure that breaks down structures and molecules into raw materials to be used for building other molecules such as proteins.

25. **Peroxisome** – membrane-enclosed structures that make and contain peroxides for digestion by importing hydrogen ions from water

26. **Phagosome** – special structure that engulfs harmful invaders.

27. **Mitochondrion** – membrane enclosed structure in eukaryotes, which produces and supplies energy bearing molecules to the cell for work by other structures. Each contains a circular segment of DNA that is independent of the nuclear DNA and can divide independently as

needed. (bacteria have similar energy producing mechanisms in their cell membranes.)

28. **Plastid** – membrane enclosed structures in plant cells that perform special functions

29. **Chloroplast** – plastid with stacks of lamella that contain chlorophyll and other structures that perform photosynthesis, to build sugars from sunlight, water and carbon dioxide.

30. **Tonoplast** – In plants, the central fluid filled vacuole that separates the plant sap from the cytosol.

31. **Cytoplasm** – entire contents of the cell exclusive of the nucleus

32. **Cytosol** – semi-fluid water based part of cytoplasm containing dissolved enzymes and salts; contains cellular structures and support system; facilitates transport.

33. **Microtubule** – hollow rod of tubulin proteins for support and structure; locomotion in flagella; aid in separation during cell division.

34. **Microfilament** – solid rod of actin protein that may also contain myosin protein for structure, support and locomotion.

35. **Intermediate Filament** – cable of supercoiled multiple strands of various proteins such as keratin for structure and support.

36. **Microtrabecular network** – network of very fine fibers in the cytosol used for structure and support.

37. **Cytoskeleton** – Mesh of microtubules, microfilaments, intermediate filaments, and trabecular network that give cells their distinct shape; used for support and structure, to hold organelles in place and for communication within the cell

38. **Flagellum** – a protein structure with moving parts that extends outside the cell and that propels the cell.

39. **Cilia** – hair-like projections from the cell membrane that aid in mobility or movement of nutrients and foreign substances. (eg, lung and trachea)

40. **Microvilli** – multiple short projections of the cell membrane that increase surface area for absorption of nutrients. (eg, intestines)

41. **Pili** – short protein projections from the bacterial wall or capsule that help to attach it to surfaces.

42. **Centriols** – two structures made of microtubules that produce spindle fibers that facilitate cell division.

43. **Centrosome** – structure that contains the two centriols and that assembles microtubules for cell division.

44. **RNA** – ribonucleic acid produced in the nucleus by replication of sections of DNA. Three base pairs of DNA/RNA code for each amino acid needed to build specific proteins. RNA codes for protein and is used to assemble them. Three types: mRNA (messenger RNA) used as protein templates; tRNA (transport RNA) used to transport amino acids to the ribosomes for assembly into proteins; rRNA (ribosomal RNA) located in ribosomes that facilitate protein production.

45. **DNA** – deoxyribonucleic acid found in the nucleus of eukaryotes or in a central (nucleoid) region of bacteria. In eukaryotes, mitochondria and some plant plastids such as chloroplasts, where present, also contain short sections of circular DNA.

46. **Gene** – a section of DNA, wound around histone proteins, that encodes for specific proteins.

47. **Chromosome** – a section of DNA containing numerous genes that is super coiled, folded and supported by proteins to form thick rope-like structures during cell division.

48. **Histones** – roughly spherical protein structures on which sections of DNA are coiled in a pearls-on-a-string arrangement.

49. **Chromatin** – DNA and its associated histone proteins appearing in a loosely coiled form between cell divisions.

50. **Extracellular matrix** – in multicellular organisms, the gel filling the space between cells that facilitates communication and adhesion to each other. In plants it is pectin; in animals it is glycoprotein (sugar-protein, eg, collagen).

CHAPTER 12

ORIGIN OF LIFE SCENARIOS

It is readily apparent that living things possess multiple levels of complexity. For even the simplest organism to survive, all of the components, whether they be physical structures or biochemicals, must perform their functions well and in concert. Living things must balance on a thin edge of interconnected complexity to survive. How is it possible to believe that all of this was built up piecemeal over millions of years, during which many of the components and functions were not fully in place, or to believe that small, stepwise changes in DNA over time result in new structures, when the incomplete sections of DNA must have existed long before there was a workable function? To believe that is not only improbable but insane! And it is not science. It is based on the Progressive philosophy that the universe is naturally progressive and will naturally, without any directions, progress from simpler to more complex and from nonliving to living. When applied to the origin of life a new principle is proposed called the Life Principle.[1] This theory assumes that the universe will naturally self-organize to produce life in any suitable environment over time.

The scenarios for the first life are equally unbelievable except to the true philosophical believer. These scenarios cannot be called theories, but at best hypotheses and at worst wild speculations. Among the speculations about where and how life emerged from nonliving matter, the most popular are as follows.

Interstellar Pre-assembly: Life self-assembled from amino acids, peptides (short sections of protein), and proteins that came to earth from space where they were assembled from stardust.

Warm Soup: In the absence of life to consume them, biochemicals that spontaneously formed accumulated in shallow seas until there were enough to create the first primitive life. This is the warm soup Darwin spoke of.

Panspermia: Life forms came to Earth from space, where they had existed for eons, thus extending the time period for their formation beyond the 4.5 billion years of Earth's existence.

Geothermal Energy: Life formed at geothermal vents that provided the energy needed to build complex biochemicals and structures that then came to life.

Deep Hot Biosphere: Life formed deep underground from hydrocarbons cooked by mantle heating to form more and more complex molecules that then came to life.

Clay Template: Life formed from biochemicals on the surface of clay, which acted as a template for assembling biochemicals and structures that eventually came to life.

Inorganic Life: Life first formed from inorganic particles such as clay, later adding organic chemicals for more efficient functions and finally rejecting or eliminating the original inorganic chemicals.

RNA World: RNA formed first and learned to make proteins and other structures through self-catalysis, later replaced by catalysis by protein enzymes.

Protein First: Proteins formed first that then assembled RNA and/or DNA and membranes.

Polycyclic Aromatic Hydrocarbons: (PAH), assumed to be abundant in space and early earth, through reactions such as oxygenation and hydroxylation, which led to formation of more complex molecules such as amino acids, proteins and RNA.

Whatever the means, it is hard to believe that all of the interlocking biochemical systems and cellular structures could have self-assembled over eons of time. The famous experiment that true believers point to as evidence of spontaneous creation of life is the Miller-Urey experiment.[2] In it, a mixture of methane, ammonia and hydrogen, which were thought to compose the Earth's early atmosphere, were subjected to an electrical spark, simulating lightening. Over time, a few of the smallest amino acids, the basic building blocks of proteins, were formed in very low concentrations within a mixture. The truth about the experiment is that it formed a tar of numerous organic chemicals often referred to a *Beilstein*, meaning a gross mixture. Beilstein is short for the largest and oldest database of organic chemicals that was first published in 1881 as *Beilstein's*

Handbook of Organic Chemistry. Its current electronic database can be found on line at Reaxys and contains many thousands of chemicals, thus the definition.

The conditions of the experiment are now not thought to have existed on the early earth. HDTKT? They took an educated guess from proxy evidence. Additionally, oxygen would have prevented many of the reactions leading to amino acids and would have destroyed many other products. However, without oxygen in the atmosphere, there would have been no ozone layer to protect the products from the destructive effects of ultraviolet rays streaming from the sun. Reaction products that were formed by lightning in the atmosphere would not be favored or exist for long enough to accumulate under such conditions. Water will also prevent or retard these reactions, and it is destructive to many products. Interfering molecules and water would have to be eliminated to create even the simplest peptide, (a short section of a protein consisting of a few amino acids hooked together by eliminating one molecule of water for each link).

The few amino acids in the experiment were formed as mixtures of right and left-handed molecules, but only left-handed amino acids are used by living things. Going from a mixture of amino acids in low concentrations in a tar containing many compounds to proteins or larger amino acids is not so evident, nor is it evident that it led to the creation of life. Forming a few amino acids in a tar in a highly controlled experiment does not point to an accidental, spontaneous creation of life or molecular evolution. If anything, it points to a designer, not the opposite. It is a leap of faith and thus not science. It is philosophy, opinion or religion, not science based on facts.

The encouraging thing about origin of life studies is that there are still multiple schools of thought, which is a healthy situation in theoretical science. The discouraging thing about origin of life studies is that we are now being pushed into the RNA world as the theory of choice. However, the jury is still out. A lot of work is being done to try to determine the best solution to the problem, but the search is far from over. Even if we can discover A route from dead chemicals to living systems, we will never know if it is THE way it occurred. It is a one-time event that cannot be fully understood by science because **Science is only concerned with predictable, repeatable and measurable aspects of the universe with which we can or could conceivably interact.**

PART 5

COSMOLOGY AS EVOLUTION WRIT LARGE

CHAPTER 13

ASTRONOMY AND COSMOLOGY

We know quite a lot about our earth, our solar system, our galaxy and our universe from observation and experiment. For example, the solar system resides in our Milky Way galaxy among many billions of stars and copious amounts of gas and dust. Our galaxy is just one of many billions of galaxies and it resides in the Local Group of galaxies which is inside the much larger Virgo Super Cluster of galaxies, all streaming toward some unseen "Great Attractor." Galaxies, arrayed in groups, are separated by millions or billions of light years[1] of empty space. The light spectra of the other galaxies are shifted to redder, lower energy, longer wavelengths; as a rule, the farther they appear to be, the greater are the redshifts. These and the many other facts confirmed by observation are not in question and represent true science.

Since science is about facts and truth, they are, by definition, scientific facts. However, the theories that claim to explain these facts are much less certain. They are based on interpretation, projection, mathematical models and sometimes pure reason (speculation) that may or may not be defined as science. It is not the purpose of this chapter to say whether these theories are correct or not, only whether they pass the test of being science, or fail and must be relegated to philosophy and belief. Remember, **science is the pursuit of truth about the predictable, repeatable and measurable aspects of the universe with which we can or could conceivably interact.** It must also be verifiable, falsifiable and testable. To be subject to science, there must be repeatability, so one-time events in the past such as the beginning of the universe do not qualify because they are not verifiable (or even observable), falsifiable or testable. We can't repeat or recreate it to test hypotheses. Furthermore, we only have one universe so no comparisons under different conditions are possible.

Whereas astronomy concerns itself more with actual observations and demonstrable or known facts, cosmology is concerned with the overall paradigm

Mathematicians deal with possible worlds, with an infinite number of logically consistent systems. Observers explore the one particular world we inhabit. Between the two stands the theorist. He studies possible worlds but only those which are compatible with the information furnished by the observers. In other words, theory attempts to segregate the minimum number of possible worlds which must include the actual world we inhabit. Then the observer, with new factual information, attempts to reduce the list still further. And so it goes, observation and theory advancing together toward a common goal of science, knowledge of the structure and behavior of the physical universe.

- Edwin Hubble, "The Problem of the Expanding Universe[3]*"*

of how the universe works and how it got here. As such, cosmology is necessarily more theoretical than astronomy. Cosmology starts with models based on one of the many mathematical solutions to Einstein's field theories,[2] and builds a picture of how they believe the universe must behave based on the mathematical solutions. It is a deductive, pure reason, approach rather than an empirical approach using observation, experiment and known facts about the real world. If observations are found that contradict these models, then the theories are modified slightly to make them fit the preconceived picture, but the model is never discarded. Please note that mathematics does not create reality; it only describes reality; but the approach of cosmology is the reverse. In this case, mathematics trumps reality. Theory should be the starting point of research, not the end product.

While astronomy uses observation and experiments of the real universe, astronomers may interpret what they find according to accepted cosmological theories. Both work within a paradigm based on increasing redshift of light with distance that is interpreted as an expanding universe, which started with a Big Bang event. The paradigm is built of a nested series of assumptions, so that new knowledge must be made to fit it. Anything outside that is not accepted as valid.

Interpretation of redshift as the speed of bodies traveling away from us in an expanding universe is at the heart of our present picture of the universe. Without interpreting the redshift as signifying an expanding universe, our whole picture collapses. Is it science or philosophy? Unfortunately, it is philosophy, based on the absence or rejection of alternative redshift interpretations and a failure to consider or fund research into the alternatives.

Could it be wrong? Yes, and it has caused many problems which have required many patches in the form of tweaked mathematical models, many adjustable parameters and unconfirmed conclusions. Problem areas include the Big Bang and its causes, the cause of the cosmic inflation and why it stopped, apparent acceleration of expansion, space itself expanding, no center of expansion, interpretation of cosmic microwave background radiation, abundance of light elements and formation of heavier elements, formation of solar systems, galaxies, clusters and super clusters of galaxies, black holes, missing mass and missing light (energy).

Before going further, please note that I have tried to keep this subject as simple as possible. To accomplish this, I have left out technical details as much as possible in describing and discussing each important point. This chapter is a long one only because I could neither abbreviate it further nor weigh the reader down with lots of heavy technical material. For those interested in more details, I suggest you look up each point of interest in introductory astronomy and cosmology books or on line.

COSMOLOGISTS TELL THE FOLLOWING STORY:

When the universe began, it all fit into a very tiny volume that then violently "exploded" and began to expand, ultimately creating all of the energy, matter, space and time. Immediately after the Big Bang when there was only very hot energy, there was an Inflationary Period that expanded faster than the speed of light due to a false vacuum with repulsive gravity, but then inflation ended. After that the universe continued to expand until it cooled enough for subatomic particles to condense out of energy. Since both matter and antimatter particles were created, most of the particles annihilated each other leaving only a small amount of leftover matter. When the universe expanded and cooled further, subatomic particles were formed into atoms.

Only when atoms of hydrogen dominated the universe did the universe become transparent to radiation, eg, light, X-rays. The very uniform cosmic microwave background radiation is the cooled remnant of the surface of last scattering, just before the universe became transparent to energy. When objects such as stars were formed that could produce ions, the neutral universe became a reionized plasma.[4] Later as bodies moved farther apart, expansion began to accelerate due to dark energy, which is a repulsive force, counteracting gravity.

Ordinary matter and energy make up less than 10 percent of the universe. Dark energy and dark matter, neither of which is directly detectable, make up the other 90-plus percent. Dark matter, which interacts only through gravity, is responsible for

1. the formation of large-scale structures,
2. galaxy rotation rates that do not decrease with distance from the center and
3. closing the universe to a finite size rather than an open universe that is infinite.

THE REDSHIFT TRAP

Edwin Hubble and others in 1929 discovered that the redshift of light from distant galaxies was proportional to the distance as calculated from apparent brightness of Cepheid variable stars within the galaxies.[5] This is called Hubble's Law and the proportionality constant is the Hubble Constant. The calculated distance depends on the size of this constant, which has changed drastically since its inception (from 500 to 70 km/sec/MPc).[6] It is important to remember that no actual extra-galactic distance has been measured; it has only been calculated from these assumptions. Because a redshift had been noted earlier in stars within our galaxy and had been attributed to movement of the source star away from us, it was natural to assume, based on Hubble's observations, that redshift of nearby galaxies was also caused by movement away from us.

This phenomenon is known as the Doppler effect and is attributed to the fact that each wave of light is emitted just a little farther away as the source recedes, thus stretching the light to longer (redder) wavelengths. Since farther is redder, farther must be faster by the Doppler effect. Because light travels at a fixed speed, taking millions of years to reach us from distant galaxies, farther is also looking back in time. For example one of our nearest neighbor galaxies, Andromeda, is calculated to be 2.5 million light years away, so we see it as it was 2.5 million years ago, assuming the distance is calculated correctly.[7]

What could explain this apparent acceleration into the distant past? Were the stars in the past moving faster than those in more recent times? At first it appeared to be so. Was the effect caused by the universe slowing

[If the redshifts are a Doppler shift]... the observations as they stand lead to the anomaly of a closed universe, curiously small and dense and, it may be added, suspiciously young. On the other hand, if redshifts are not Doppler effects, these anomalies disappear and the region observed appears as a small, homogeneous, but insignificant portion of a universe extended indefinitely both in space and time

- Edwin Hubble, Roy. Astron. Soc. M. N., 17, 506, 1937

down with time? If the expansion is slowing down, could it eventually stop and then start to contract? Instead, almost from the beginning, due to preconceived mathematically based theories postulating a beginning from a much smaller size, the redshift was seen as an expansion of the universe, not as contracting or slowing. But what could explain the acceleration into the past?

Hubble was never sure that recessional speed was responsible for the redshift of galaxies, but he knew of (or accepted) nothing else that could explain it. In later years, he speculated about the intergalactic medium interacting with the light, rather than expansion, as the cause of the redshift. He is credited with discovering the expanding universe and thus the Big Bang, but after his earlier work, he spent the rest of his life working to refute it.

REVIEW OF HUBBLE'S LATER EVIDENCE THAT REDSHIFT IS NOT SPEED RELATED

Source: "The Problem of the Expanding Universe," Edwin Hubble, *American Scientist*, Vol. 30, April 1942, No. 2

1. Light from stationary objects dims in proportion to the square of the distance.

2. Light from receding objects also dims in proportion to recessional speed of the source expressed as a fraction of the speed of light. (Speed that is 10 percent of the speed of light produces a 10 percent dimming because fewer photons per unit of time reach the observer.)

3. A plot of redshift versus dimming of stationary objects from distance is linear (produces a straight line).

4. A plot of redshift versus dimming from both distance and recessional speed is not linear.

5. The cosmological principle assumption of cosmology theory states that the universe is uniform and isotropic (the same everywhere) so that an observer will see the same thing no matter where they are in the universe.

6. Surveys of galaxy populations, assuming only dimming from distance, produce uniform populations at various distances (out to an estimated 500 million light years in this case).

7. Surveys of galaxy populations, assuming dimming from both distance and recessional speed, produce increasing populations into the past. This violates the cosmological principle of an isotropic universe.

8. Dimming from recessional speed would make objects nearer than they appear (due to extra dimming).

9. If redshift is assumed to be from recessional speed, the universe is smaller, more compact and younger than it would be if speed were not a factor.

10. Redshift increases with distance and time so if it is Doppler effect (recessional speed), then the supposed expansion appears to be slowing down. (Note that cosmologists solved this problem by assuming expanding space and a cumulative effect of expansion of space itself between objects.)

11. Although he does not rule out an expanding universe, he favors an expansion so slow that it will not be detectable apart from redshift. Quote: "We may still suppose that the universe is either expanding or contracting, but at a rate so slow that it cannot now be disentangled from the gross effect of the superposed redshifts."

12. Although he contends that redshift is not due to recessional speed, ie, Doppler effect, he does not identify a cause in this paper. Quote: "Meanwhile, on the basis of the evidence now available, apparent discrepancies between theory and observation must be recognized. A choice is presented, as once before in the days of Copernicus, between a strangely small, finite universe and a sensibly infinite universe plus a **new principle of nature.**" (emphasis added)

CONDITIONS:

1. Hubble never mentions or invokes expanding space.

2. Although he granted that space may be curved from mass within it, as predicted by the theory of relativity, he calculated that the total mass must be much greater than the visible matter.

3. In this paper, Hubble excluded the local group of galaxies, recognizing that their motions were not typical. He asserted that, in an expanding universe, groups would maintain their orientation.

4. Surveys of galaxy populations were done with 100-inch Mt. Wilson and smaller telescopes. The larger the telescope, the farther it can see, which he estimated to be 500 million light years and less.

5. His estimate of the recessional acceleration (he never called it the Hubble Constant), if Doppler speed is the cause of redshift, was 100 miles/second/light year (49.3 km/sec/MPc). Present estimate is 70 km/sec/MPc. His original theory estimated it at 500 km/sec/MPc.

6. Redshift was calculated out to an estimated 250 million light years and, if due to the Doppler effect, the speed would be 25000 miles/sec or 1/7 the speed of light. (the distance is roughly 2 percent of 13.7 billion light years, the theoretical time since the Big Bang.)

In formulating his original theory, Hubble worked with nearby galaxies whose redshifts had a value much less than one, and he verified their distances by comparing the apparent and actual known (assumed) brightness of Cepheid variable stars within them.[8] According to the original equations of the theory, a redshift (z) of 1 is equal to the speed of light since it is defined as a ratio of the actual speed (v) to the speed of light (c) or ($z = v/c$).

This was a nice picture until some years later galaxies were found that approached or even exceeded a redshift of one. According to Einstein's relativity equations, light speed in a vacuum is thought to be the maximum speed possible, so this was a problem. At the present time, distant galaxies have been found that exceed a redshift of eight. Obviously, something besides the recessional speed of the source was causing these strange observations. Many possible explanations were proposed, but the theory that won out, politically, because it resonated with the dominant theoretical cosmologists of the day, was that the actual space between the galaxies was uniformly expanding causing a cumulative effect for farther galaxies carried within it. To explain acceleration into the past and speeds apparently faster than the speed of light, cosmologists assumed that the very space between galaxies was expanding, thus a cosmological redshift was added to pure Doppler redshift of Hubble's original theory.

This neatly fit a comfortable picture of a consistently expanding (progressively evolving) universe from the time of its beginning in a Big Bang event that was first predicted by Alexander Friedman and Georges Lemaitre from their independently derived solutions of Einstein's gravitational field equations, and later championed by George Gamow and others. Later this was modified to include an original inflationary period just after the Big Bang event and a more recent acceleration of expansion.

It should be noted that the present redshift with distance theory excludes the nearby galaxies because they do not conform to the picture due to local gravitational effects. As an example, the Andromeda galaxy is actually blue shifted (coming toward us) and it is expected to collide with our galaxy in about 4.5 billion years. What is interesting is that in his original work Hubble only worked with nearby galaxies. He also mistakenly used some non-Cepheid variable stars so that some of his distance estimates were wrong. His original calculations of a linear relationship between apparent brightness and redshift were only valid if objects were assumed to be fixed, not receding.

Therefore, the original premise, that Doppler (recessional speed) redshift is directly related to distance of nearby objects, has been invalidated and has been changed to fit the picture of a Big Bang with expanding space combining Doppler redshift with cosmological redshift. Since redshift can and does exceed 1, meaning that the objects are receding faster than the speed of light, the relationship of redshift to distance is no longer linear as originally proposed, but is described by a more complicated equation.

A competing theory to the Big Bang was an expanding space Steady State model by Fred Hoyle and others in which new matter was continually being formed by vacuum fluctuations to fill the spaces so formed, with no mention of an origin of the universe. Hoyle's team never described how such new matter was being formed. Both theories depend on assumptions based on the redshift as a Doppler effect. Another theory, rejected earlier but being re-examined by some astronomers today as an alternative, is a truly infinite, equilibrium static universe first proposed by Walther Nernst in 1928 and supported by many other famous physicists.[9] In this theory, intergalactic redshift is not interpreted as a Doppler effect, but as caused by other factors such as gravitational drag and refraction caused by varying density of matter in intergalactic space.

EXPANDING SPACE-TIME

According to the Standard Model, space is assumed to be a vacuum more complete than any we can produce on earth. Its estimated density is one hydrogen atom per four cubic meters, and that includes all the material in the galaxies averaged. Please note that this is a calculation based on assumptions that cannot be verified. It is therefore a belief or guess, not a scientific fact. Is space a something or a nothing? Could space itself really expand? Can nothing expand? Is this science or philosophy? Unfortunately, it is the latter. It is a belief that fits a Progressive picture and sidesteps the thorny problem of speeds apparently faster than light, while retaining the Doppler effect, enhanced by expanding space, as the only or at least the dominant cause for redshifts.

By this scheme, galaxies could all be receding at the same rate, but the space expanding between them, by a cumulative effect, makes them only appear to be moving faster than light. Like raisins in a rising loaf of bread, the distance between each adjacent raisin is expanding at the same rate, but the farther apart they are, the greater the actual distance increase because it is a combination of the expansion of all of the spaces between each intervening raisin. This picture of expanding space also assumes that the gravity within galaxies, and even galaxy clusters, decouples them from the expansion. Otherwise, matter could just disperse as if by evaporation.

> But it is always safe and philosophic to distinguish, as much as is in our power, fact from theory; the experience of past ages is sufficient to show us the wisdom of such a course; and considering the constant **tendency of the mind to rest on an assumption** and, when it answers every present purpose, **to forget that it is an assumption**, we ought to remember that it, in such cases, becomes a prejudice and inevitably interferes, more or less, with a clear-sighted judgment.
>
> I cannot doubt but that he who, as a wise philosopher, has most power of penetrating the secrets of nature, and guessing by hypothesis at her mode of working, will also be most careful, for his own safe progress and that of others, to **distinguish that knowledge which consists of assumption, by which I mean theory and hypotheses, from that which is the knowledge of facts and laws**; never raising the former to the dignity or authority of the latter, nor confusing the latter more than is inevitable with the former.
>
> *- Michael Faraday, Experimental Researches in Electricity, Vol. 2, p. 286, Chapter on Electric Conduction. (Emphasis added.)*

Einstein replaced the earlier assumed space-filling aether, which had never been detected or verified, with space-time. Space (or space-time) has been treated as a something ever since, although there is no way to confirm its material existence any more than that of the aether. One of the reasons why aether was needed in the first place was the seeming paradoxes of gravity and the propagation of waves without a carrier for them. Let me explain.

GRAVITY

Gravity obviously acts over vast distances to hold and draw masses together into stars, solar systems, galaxies and clusters. Why is there gravity and how does it work? The mechanism for this is poorly understood, although its influence can be calculated precisely by celestial mechanics as first proposed by Newton.[10] There is still a debate about whether gravity travels at some finite speed or is instantaneous, and also whether it has a limiting distance of influence. We do know that light takes 8 minutes to reach earth from the sun but the sun's gravitational pull, as seen in its contribution to tides, occurs where the sun actually is, not where we see the sun to be (as it was 8 minutes ago). Is gravity instantaneous or at some speed much greater than the speed of light? Greater than light? Isn't light speed believed to be the upper limit?

What could possibly explain this seemingly instantaneous effect over vast distances? It was thought that there must be some substance that carries gravity from one body to another, and thus the aether. With Einstein's space-time, the problem was again assumed to be solved. Space-time (ie, aether) must carry the gravitational influence as a warping of space-time by mass, like a bowling ball compressing a mattress but in all three dimensions. Bodies would then naturally fall into this "gravity well." Thus, gravitational influence would be instantaneous and at the same time decrease with the square of the distance. In reality we are back to aether, only by another name backed by new theoretical calculations based on assumptions. The warping of space-time can be useful as an analogy in instruction to see the law of gravity, but is it real or just a great thought picture that illustrates its effects? Could it be a field like electromagnetism?

PROPAGATION OF WAVES

Since space is thought to be an almost perfect vacuum, (empty), how is light propagated without something to carry it? We know by experience that radiation across the entire spectrum[11] will travel through a vacuum in the laboratory, but how is it possible? All radiation is thought to be electromagnetic waves and is classified as a "standing wave." So, in a vacuum, or space, what is being waved? How can it continue to travel in nothingness for vast eons of time apart from its original source which may or may not continue to exist in its original form or location? Here again, a space-filling aether or space-time must be carrying it.

Secondly, light (any radiation) has been redefined not as a wave but as a particle or packet of radiation, the photon. Actually, it is treated as both a wave and a particle as needed. As a particle, once set in motion, it can continue indefinitely without attachment to its source or needing a carrier. All matter constantly emits radiation in the form of light, heat, et cetera, depending on its temperature, from very short wavelengths (high frequency)[12] gamma rays to very long wavelengths (low frequency) microwave or radio waves - the lower the temperature, the longer the wavelength. Whereas high energy events like fusion emit high energy gamma rays, cold dark matter in space would emit in the low energy microwave range.

ENERGY AND MATTER FACTS:

Except for processes like fusion that change matter and/or antimatter into energy, radioactive decay and exothermic[13] chemical reactions, atoms emit radiation that has been absorbed from their surroundings. Any interaction between matter and energy results in a net loss of energy. For visible and UV light, the energy that is absorbed by an atom kicks an electron up to a higher (excited) energy orbital, and it is re-emitted when the electron falls back to a lower or its original (stable) orbital. In this process, there is a net loss of energy resulting in longer wavelength (redder) emission. Fluorescence and phosphorescence are examples of this. Similarly, lower energy microwaves move (translate) molecules and infrared light excites molecular bonds to vibrate faster. Even in high energy encounters, such as matter with gamma or X-ray energy, that splits (ionizes) atoms and molecules, there is a net loss of power, resulting in lower energy, longer wavelength (redder) emissions.

DOPPLER EFFECT AND OTHER THEORIES

All of astronomy and cosmology are tied to the picture of an expanding universe by the Doppler redshift paradigm and its seemingly paradoxical speeds faster than light. It was necessary to invent things like expanding space to explain it. Can it all be true? Possibly. Can there be other possible explanations that don't require such gymnastics? Could there be multiple causes? Yes, there could be other and perhaps multiple factors. All we really know is that redshift seems to increase with distance.

Politically, the Big Bang and its expanding space based on the Doppler effect won out in the cosmology community, but there were and are other explanations or at least contributing factors that might be involved. Other theories have been maligned, ridiculed and misrepresented in order to hang onto the standard line of Doppler effect with expansion of space as the only or the major cause. Other scientific theories have been systematically excluded from scientific publications and funding for research. Reality has been distorted, reinterpreted and ignored in order to fit it to a cosmology based on a mathematical model that itself was based on deductive or pure reasoning; in other words, an assumption. There are many "anomalies" that just don't fit the model and Just-so stories have been published as facts or at least well founded theories.

Any redshift mechanism must shift an entire spectrum, including its family of elemental absorption and emission lines, as a unit and not just one wavelength at a time or different rates for different regions. It must also not scatter or block the light from distant objects thus blurring or obscuring them from view. Those restrictions would rule out most collision type phenomena due to scattering or wavelength selectivity. Additionally, fluorescence, phosphorescence or other atomic re-emission of absorbed light would not be candidates because they would apply only to specific wavelengths of light and would not preserve the entire spectrum. Some sort of refraction might fit, but only if the refraction is perfect enough to preserve the spectrum, complete with the original absorption and emission lines. Refraction occurs when light travels from one density of material to another. The light is bent from a straight path by an angle determined by the differences in relative densities of the media and may be wavelength specific.

In order to see the increasing redshift with distance, if refraction is the cause, there must be repeated perfect refraction events all along the way.

> Supporters of the big bang theory may retort that these [alternative] theories do not explain every cosmological observation. But that is scarcely surprising, as their development has been severely hampered by a complete lack of funding. Indeed, such questions and alternatives cannot even now be freely discussed and examined. An open exchange of ideas is lacking in most main- stream conferences. Whereas Richard Feynman could say that "science is the culture of doubt", in cosmology today **doubt and dissent are not tolerated**, and young scientists learn to remain silent if they have something negative to say about the standard big bang model. Those who doubt the big bang fear that saying so will cost them their funding.
>
> *- Open letter from the international Alternative Cosmology Group,*
> *Published in New Scientist, May 22, 2004.*
>
> *ACG website: www.cosmology.info (Emphasis added)*

Let's look at one of the other theories that was and is perhaps the greatest threat to the present paradigm. The so-called "tired light" theory[14] of Fritz Zwicky is sometimes mentioned, but it is usually mischaracterized as resulting from collisions with high energy electrons in space, the Compton effect. Going back to Zwicky's original paper[15] the true theory can be shown to specifically eliminate the mechanisms cited in the mischaracterizations as causing too much scattering of light to allow us to view distant objects. The popular story line is that his theory involves light repeatedly hitting and rebounding from fast moving free electrons in space, losing some energy with each encounter, and thus moving to longer and longer wavelengths, ie, becoming redder with distance. The more collisions and thus the farther the distance, the redder the light would become. However, as stated above, this mechanism would cause too much scattering to observe the distant galaxies clearly and would not preserve the entire spectrum well.

Note that the Compton effect could explain redshifts between the center and limb (edge) of the sun, and in solar flare activity, due to free electrons in the sun's atmosphere and a difference in the light path through the sun's atmosphere. In this case, the blurring from scattering would not be a significant hindrance to observation. However, the heavier the solar element, the greater the redshift. It is not a coordinated shift of the spectrum in its entirety.

Although Zwicky reviewed other possible explanations involving collisions, he really theorized that the drag of gravity, without collisions, was the most probable reason for loss of energy by photons of light, thus moving them to longer, redder wavelengths. Einstein theorized that light would be affected by gravity as it passed other objects. This has seemingly been demonstrated by viewing stars behind the sun during an eclipse as their light is bent by the sun's gravity.[16]

According to Zwicky's theory, each time light passes an object in space it gives up some of its energy to that object through gravitational interaction, becoming redder. Thus, farther is redder because of more gravitational encounters along the way. Alternatively, by this theory, if gravity has no distance limits, light may be affected by the gravity of the universe as a whole, so that the longer the time that the light travels, the longer it is affected by gravity, thus becoming redder with time. Remember that Hubble calculated a linear distance to redshift relationship that only fit a static universe. His comparison of results calculated for either an expanding or a static universe, based on expected dimming of light with distance, convinced him that the redshift relationship could not be merely a Doppler effect.[17]

What are some of the other possible explanations or contributors if there are multiple factors? See the brief list below which, by no means, exhausts the theories that are out there.

REDSHIFT THEORIES:

Doppler effect (Pure or Classic) – speed of the source away from the observer causes a shift to longer (redder) wavelengths caused by the source being farther away each time a photon (or wave) is emitted, effectively "stretching" the wavelength.

Cosmological Doppler effect (Expansion of Space) – speed of the source appears to be faster the farther back in time it is, because of the expansion of the very space in which it is embedded. Doppler effect is magnified by the cumulative effect of uniformly expanding space.

Relativistic Time Dilation – when a source is moving at high speeds, time on the object is (or appears to be) passing slower to an observer outside its frame of reference, thus causing the light to appear redshifted. Distant supernovae type 1a intensity curves with longer decay times than expected are cited as evidence of time dilation.

Gravity retardation – a source of light that is a massive object could produce redshifts due to its extreme gravity retarding light emitted from it. This would be source specific only, and would not apply to all objects at the same distance. Quasars, with their large redshifts could be an example of this and thus may be closer than their calculated distances.

Gravitational Drag (Zwicky) – each time a light wave passes a mass in space it gives up some of its energy to the mass by gravitational interaction, thus becoming redder due to loss of energy. Therefore the redshift would be distance, not speed, dependent.

Modified Gravitational Drag – the universe as a whole exerts gravitational retardation or drag on light so that the longer this force acts the more redshifted the light becomes. Thus, redshift is correlated to distance (and time), not speed of recession.

Plasma Field Retardation – plasma field, assumed by this theory to fill space, stretches light each time a plasma particle is passed and the results are cumulative. It also reduces the intensity, and blurs the image very slightly with each interaction.

Compton Effect – each collision with high speed electrons, assumed by this theory to fill space, would reduce the energy of the resulting rebounding light resulting in reddening of the light. This has two major problems. The first is scattering of the light so that blurring would approach occlusion with distance. The second problem is that specific wavelengths, not the entire spectrum would be involved with electrons having specific energies by the quantum effect.

It is interesting that, although Zwicky's theory of redshift through gravitational drag (1929) has been rejected politically by the cosmological community because it did not fit with their preferred paradigm of an expanding universe, other theories proposed by Zwicky that were originally rejected have later been confirmed or at least adopted. Among these are supernovae as the source of cosmic rays,[18] neutron stars or black holes resulting from supernovae collapse,[19] formation of galaxy clusters[20] dark matter (1933) to explain galaxy and cluster rotation rates, gravitational lensing[21] and supernovae (type 1a) of consistent brightness that can be used as standard candles. Zwicky's catalogue of galaxies[22] is still being updated and used today. He also designed some of the earliest jet engines.[23]

PROBLEMS WITH THE BIG BANG

The origin of the universe is a one-time event that cannot be repeated, tested, falsified, or even observed. It is closer to the creation myths of aboriginal peoples than to hard science, regardless of the extremely complex mathematical calculations used to justify and fortify it. Proven, in the sense of a mathematical proof, doesn't make it true in the real universe. The Big Bang theory rests on an original assumption to which mathematics has been fitted, massaged, misused and tortured.

The mathematics are elegant and awe-inspiring, but cannot be said to represent reality unless experiments and observations back up each assumption and conclusion. This is not the case. The mathematics are filled with many adjustable parameters, each intended to solve a different problem. It is not a single set of formulas that fits all cases, but a hodge-podge of fudge factors to make it fit each new problem. Many exotic and unobserved phenomena have been invented because anomalies have required new explanations. Among these are singularity at the beginning, dark energy (as a cause of expansion and acceleration of expansion), dark matter (to explain large scale and galaxy gravitational effects), black holes, primordial inflation with false vacuum and repulsive gravity just after the Big Bang, acceleration of expansion later, expansion of space, vacuum energy and virtual particles, multiple dimensions and multiverses. Among the unanswered questions are the origin of heavier elements, the origin of the galaxies, stars and large scale structures and there is always the question of the first cause or uncaused cause.

> However, the most unhealthy aspect of cosmology is its unspoken parallel with religion. Both deal with big but probably unanswerable questions. The rapt audience, the media exposure, the big book-sale, tempt priests and rogues, as well as the gullible, like no other subject in science.
>
> *- Michael Disney*[24]

SINGULARITY

According to the standard model, the universe, including all matter, energy, space and time, began with a gigantic "explosion" from a singularity. Note that the operative word here is "model", in the sense of a mathematical model. A singularity is a theoretical point with no dimensions, and cannot exist in the real world. For anything to exist, it must have height, width and depth dimensions. In other fields of mathematics, either a singularity or an infinity result means the equations failed, and either the theory or the equations based on it must have some flaw.

Steven Hawking and Alan Guth were so concerned with this that they built a mathematical model that attempted to bypass a singularity in favor of a pouch that curves in on itself, sort of like the toe of a sock. This inflationary period "solved" another problem. The horizon problem basically is the question of why the universe is very uniform, as defined by the cosmic microwave background, even in areas on opposite sides of the universe that have not been in communication since early in the universe. See next section. Some cosmologists have replaced the singularity in favor of something very, very tiny although their own calculations lead to a singularity. Is it science? No, it is a mathematically fortified philosophy. The mathematical equations are based on deductive reasoning, not on reality, experiments or observations. Remember that Einstein's field equations could have many possible solutions, depending on the assumptions and parameter values chosen.

INFLATIONARY PERIOD

Because the universe, as defined by the cosmic microwave background, appears to be uniform in all directions, there is a horizon problem. In a universe that is less than 20 billion years old, how can light have traveled from one side to the other, a diameter of approximately twice the age. How can structures be so similar when they have not been in communication for most of their existence? The inflationary period just after the Big Bang was introduced to smooth out the universe and fix the horizon problem.

If all areas started out very, very homogeneously before matter condensed from pure energy, then that could explain the uniformity of the known universe. However, to invent this inflation of space, it was necessary to postulate unknown and unknowable causes and effects that do not exist today. According to the theory, a false vacuum was formed by the early expansion causing gravity to be repulsive instead of the consistently attractive gravity we know today. As such, in a universe containing only energy, expansion could inflate faster than the speed of light. Yes, you heard that right. False vacuum, repulsive gravity and speeds faster than light are completely made-up properties and can never be detected, observed, tested or falsified. They have little value beyond myth that solves a problem with cosmology.

COSMIC MICROWAVE BACKGROUND

The cosmic background temperature, ie, cosmic microwave background, CMB, interpreted by cosmologists as the extremely redshifted afterglow of the hot Big Bang (or at least the surface of last scattering),[25] is thought to be the best evidence for the expanding universe theory. However, the temperature of thinly dispersed matter in space as a result of residual starlight was earlier calculated and predicted by Guillaume, (5 K < T < 6 K),[26] Eddington, (T = 3.1 K), Regener and Nernst, (T = 2.8 K), McKellar and Herzberg, (T = 2.3 K) Finlay-Freundlich and Max Born, (1.9 K < T < 6.0 K), based on a universe in dynamical equilibrium without expansion.[27]

After Penzias and Wilson experimentally found the cosmic microwave background radiation to be consistent with a temperature of 2.7 K, Gamow, who had claimed to be the originator of the Big Bang Theory, also erroneously claimed he had been the first to predict the background temperature and claimed the result as evidence for the Big Bang. However, his estimate was not only not the first, but was 7 K with an upper limit of 50K. Thus, the "best evidence" or "smoking gun" for the Big Bang is called into question, and is only valid if the Doppler effect of an expanding universe from a hot Big Bang beginning is assumed a priori,[28] not the residual temperature due to starlight on dispersed matter. This is circular reasoning and is a logical fallacy.

Note that the cosmic background radiation (DEBRA or Diffuse Extragalactic Background radiation) includes all wavelengths from gamma rays through radio waves. See diagram below.[29] The power of radiation is inversely related to the wavelength so that shorter wavelengths are more powerful. In the diagram, the power is multiplied by the wavelength, so the resulting graph shows the relative contribution at each wavelength. As you can see, although the CMB is the highest, other regions contribute significantly. (Note that the power scale is logarithmic so each division on the graph is ten times that of the preceding one.)

In addition to these radiation fields, magnetic fields and several types of particles also make up an interstellar and intergalactic background. Among these are cosmic ray background, cosmic neutrino background and various magnetic fields. Cosmic rays are energetic bare atomic nuclei from beyond earth, composed of 87 percent protons, 12 percent helium nuclei and a small amount of heavier nuclei as well as high energy electrons. They come from all directions and tests indicate that the flux has not changed in millions of years. Cosmic rays have a wide range of energies because they originate from multiple sources, from the solar wind to distant galaxies. The local neutrino background comes from the huge numbers of neutrinos streaming from the sun as a product of fusion reactions.[30] Because of their small size and electrical neutrality, most neutrinos can pass through stars and planets unimpeded.

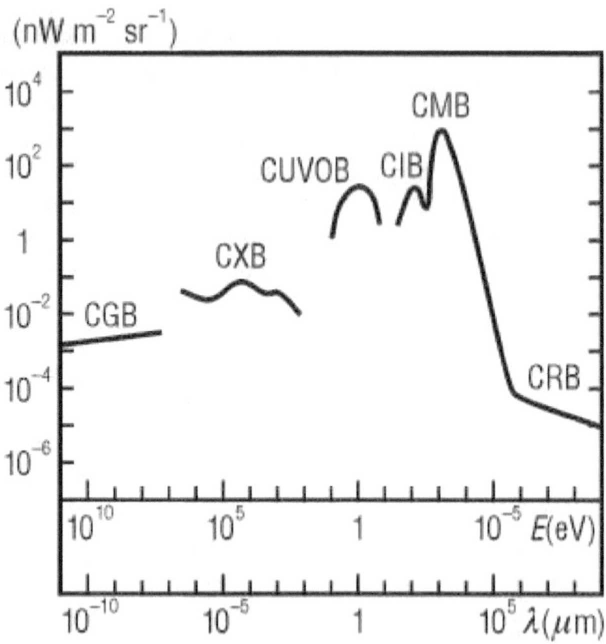

Figure 1. Spectral Radiance vs. Wavelength (λ) in microns (μm)

This shows a schematic picture, based on many different data sets, of the spectral intensity (also called spectral radiance) multiplied by wavelength of the DEBRA over all the electromagnetic spectrum. See key to abbreviations below.

CGB = Cosmic Gamma Ray Background
CXB = Cosmic X-Ray Background
CUVOB = Cosmic UV-Visible Background
CIB = Cosmic Infrared Background
CMB = Cosmic Microwave Background
CRB = Cosmic Radio Wave Background

All stars emit neutrinos in the same way, so there must be an interstellar and intergalactic Neutrino Background. Thus far, cosmic neutrinos have only been detected from the sun and a nearby supernova (SN 1987A in the Large Magellanic Cloud), possibly because the extra-solar neutrinos are believed to have much lower energy than can be detected by present instruments. Like the cosmic microwave background, the (assumed) cosmic neutrino background is attributed by cosmologists to events shortly after the Big Bang, so that their assumed lower energy is explained by expanding space. This does not explain the failure to detect the neutrinos from nearby stars. Incidentally, intergalactic neutrinos are one of the candidates for the supposed 90 percent missing mass of the universe, termed "dark matter." On that same subject, the emptiness of space may be another unsupported assumption or belief. (See section on dark matter.)

As you can see, intergalactic space is not so empty after all. All of these rays and moving particles must surely result in magnetic fields. Star systems, galaxies and groups of galaxies are believed to have magnetic properties that affect their forms and behavior. Magnetic fields on a cosmic scale are poorly understood and are in need of further research. LOFAR, Low Frequency Array, radio telescope in the Netherlands and SKA, Square Kilometre Array radio telescope, scheduled to be built in Australia, are two facilities that propose to investigate cosmic magnetic fields.

DARK MATTER AND DARK ENERGY

Although dark matter and dark energy have never been directly detected, they are thought to be necessary to the standard cosmological model. This is a case of ad hoc (special) theories added to the mathematical model to make it fit observation more closely. Rather than consider that their model might be wrong or incomplete, cosmologists have invented undetectable phenomena to force a fit. Are they real? Maybe, but they are definitely not science, they are philosophy because they are not observable, not testable and not falsifiable.

DARK MATTER

Dark matter is needed to close the universe in the standard expanding universe model, to form galaxy clusters and walls, and to account for orbital speeds at the outer edges of galaxies that are greater than the galaxy's assumed mass can account for. Some calculations indicate that the observable matter in the universe accounts for only a few percent of the total; the balance is thought to be dark matter. This missing mass could be exotic or ordinary matter. The two most favored candidates are WIMPs and MACHOs. WIMPs are exotic Weakly Interacting Massive Particles. MACHOs are Massive Compact Halo Objects made up of ordinary, non-luminous matter such as neutrinos, planets, brown dwarf stars and black holes.

Neutrinos are one of the candidates for unseen matter. The sun and other stars are constantly streaming huge amounts of neutrinos created in the fusion reactions that power stars. It is estimated that a trillion neutrinos each second are impinging on every square meter of earth's surface. Since these tiny neutral particles rarely interact with ordinary matter, most of them pass harmlessly through the earth. Only a few hit heavier elements, affecting radioactive decay rates. During solar flares there appears to be a dip in the rate of radioactive decay of laboratory materials. There is also a daily and seasonal fluctuation reflecting orbital distance differences.[31]

If all stars are constantly creating neutrinos and streaming them throughout the universe, then there must be a tremendous amount of mass from neutrinos between and within all of the galaxies. It is assumed that there is a cosmic neutrino background (CNB) which cosmologists assume comes from the Big Bang, like the CMB (cosmic microwave background). None of these intergalactic neutrinos has yet been detected. It is assumed that they are much less energetic than solar neutrinos so that they are undetectable by present instruments because, like CMB, they are assumed to be far redshifted remnants of the Big Bang. What about the neutrinos from nearby stars? Surely, these have not lost enough of their energy to be undetectable. We should be awash in interstellar and intergalactic neutrinos from nearby objects.[32]

ACCELERATION OF EXPANSION

Dark energy is needed to account for acceleration of the expansion of the universe. It is thought to represent a net repulsive force counter to gravity's attractive force. The acceleration of expansion was first proposed because distant type 1a supernovae appear dimmer than expected at their redshift distance. At these distances, the redshift to distance relation is no longer linear, implying a change in rate of expansion, ie, acceleration. This assumes there is no attenuating matter between the object and the observer. It also assumes that type 1a supernovae are a reliable standard candle properly calibrated with the previous series of standard candles[33] each of which is more uncertain than the last, often with little or no overlap. Since this is the only one that shows a different rate, resulting in a quasi-exponential rate, there is some uncertainty about its interpretation.

The theory is that without this repulsive force counteracting gravity, the universe would have collapsed in on itself. There are two main competing candidates for dark energy: the cosmological constant and quintessence. The (new) cosmological constant is thought to be present everywhere equally and is unchanging. It is equivalent to vacuum energy, which is a concept borrowed from particle (quantum) physics, that every point in space has its own energy that can spawn virtual particle-antiparticle pairs, which immediately self-annihilate producing energy. Quintessence[34] is a hypothetical energy field filling the universe that is changeable depending on the mass of objects within it. The assumption is that, as the universe expanded, it became repulsive about 10 billion years ago, starting acceleration of expansion.

Detection of dark energy has never been confirmed[35] and the whole premise of acceleration of expansion is based on interpretation of data that is itself based on an assumption of no attenuation of light and that type 1a supernovae are all the same intrinsic (real) brightness. Since matter and energy are equivalent by the theory of relativity, dark energy is thought to have gravitational effects that contribute to the overall mass-energy of the universe.

GALACTIC ROTATION RATES

There is definitely something causing the stars in the outer reaches of galaxies to revolve at much higher rates than would be expected by classical celestial mechanics. The outer planets of our solar system move ever slower the farther they are from the sun in order to maintain the balance between the sun's gravitational pull and centrifugal forces from orbital speeds. This is not the case for galaxies where the speeds of outer rim stars are closer to the speeds of inner arm stars, so that galaxies appear to rotate almost in unison like a giant wagon wheel. Surely there is something needed to account for this. Either our understanding of celestial mechanics at these large scales is lacking, or there really is something massive causing it.

One possibility is that prior mergers have given these stars extra momentum, but that might result in these stars escaping the galaxy unless another force stabilizes their positions. Another possibility is that the gravity of nearby stars may have greater influence on outer arm stars than the distant galaxy core. Since the effects of gravity decrease with the square of the distance, inner arm stars would feel greater gravitational pull from the center than outer stars. This is basically why the outer stars are expected to revolve slower to balance gravitational centripetal (inward) and orbital centrifugal (outward) forces in a similar way to our solar system.

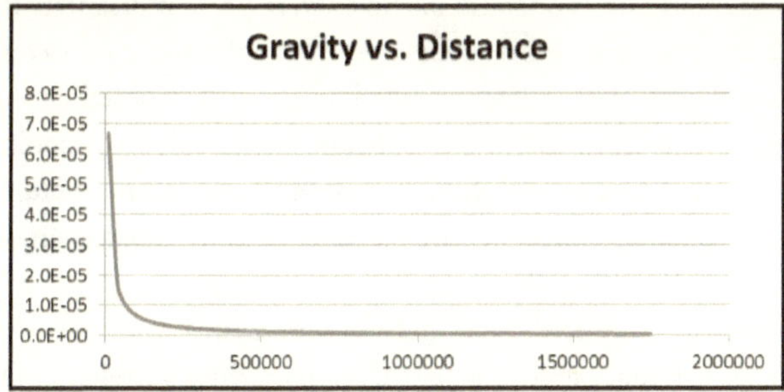

These gravitational anomalies are often treated as though this is a two body problem, not a multiple body problem. In celestial mechanics, the orbits of two bodies are easily calculated, each orbiting their mutual center of gravity. In the case of the sun and earth, the difference in mass is so great that this center of gravity is so near the sun that it is barely noticeable. When you add a third body, eg, Earth, moon and sun, the system becomes more chaotic and the orbits can only be approximated. Over extended periods of time, the incremental orbital uncertainty results in greater and greater uncertainty. With thousands of nearby stars and a galactic center involved, the situation is much more complicated. These calculations for galaxies often only consider the influence from the mass inside the orbits of stars, not from the totality of all bodies within and outside the orbits in question.

What if the gravitational influence of the galaxy core decreases with distance so much that the gravity of nearby communities of stars have greater influence than the core? In that case, tidal forces might drag the outer arm stars along at speeds similar to inner arm stars and stabilize their positions. In this case, MACHOs and WIMPs may not be necessary.

In addition to gravity, a galactic electromagnetic field may contribute to the observed nearly synchronous orbital speeds. If the entire galaxy is embedded in an electromagnetic field supported by a plasma of interstellar material that could help to unify movement within it. Plasma is an ionized gas that is the fourth state of matter, along with solid, liquid and gas. Plasma has unique electrical and magnetic properties. The earth and planets are embedded in a magnetic field created by the sun and the solar wind, made up of plasma, called the heliosphere. If the galaxy acts like a gigantic magnetic system, then nearly synchronous orbital speeds may be the result. The truth is that we do not understand this phenomenon and it is OK to say, "I don't know" until real discoveries, not mathematical models or speculations, show us the answers.

ORIGIN OF HEAVIER ELEMENTS

According to the standard model, no particles were present just after the Big Bang event because the energy was too intense. Rather, subatomic particles condensed later from pure energy after space expanded and thereby cooled enough for their formation. Here, it is assumed that expansion is equivalent to cooling. At first, according to the standard model, both matter and antimatter were created, and then annihilated each other, leaving only a small amount of excess matter to form the atoms of the universe.

In this scenario, only hydrogen, helium and a tiny amount of lithium were created out of the original subatomic particles. According to stellar nucleosynthesis theory, all of the other elements were created inside stars by fusion (helium through iron) or through the deaths of stars in supernovae explosions (for heavier elements). Only second or later generations of stars, which were formed from the remains of supernovae, would contain any heavier elements. In the language of astronomy, they would acquire metalicity, ie, heavier elements, in subsequent generations.

According to the standard model, the universe is 13.7 billion years old, and we have viewed galaxies as far back as 95 percent of the way to the Big Bang. These early galaxies show no distinct signs of evolving from a more primitive state, but appear to be every bit as complete and organized as later, nearer galaxies, including metalicity where it is possible to measure it. This is a mystery. How could they be that well developed and be composed of second or later generation stars at that early stage? It seems likely that either 1.) the age of the universe is wrong or 2.) we are not looking as far into the past as we think we are, or 3.) the theory that nucleosynthesis of heavier elements only occurs inside stars or supernovae is wrong. This latter does not mean that stellar and supernovae nucleosynthesis does not occur; only that it may not be the only source of heavier elements. Similarly, the theory of evolution of galaxies is also in question. How could completely evolved galaxies exist that early in the history of the universe?

One theory to explain this absence of primitive galaxies and low or no metalicity stars proposes that the earliest stars with no metalicity were huge, unstable and short lived, spraying heavier elements throughout the universe through supernovae explosions. However, if they were too large, they would have collapsed into black holes. None of these pre-metalicity galaxies have been detected, so they either never existed or had lived and died very early in the universe's history.

Within our own galaxy, gas clouds, which are mostly composed of hydrogen, are known to be poor in heavier elements, whereas dust clouds are enriched in heavier elements. Stars within our galaxy that are nearer the galactic center have higher metalicity than stars farther out, supposedly because of higher rates of supernova formation near the crowded and more active center. Stars in nearby globular clusters have lower metalicity overall than the Milky Way to which they are satellites. The reason for this is unclear, but could be accounted for if we assume that the gravity of galaxies draws in material, and speeds up or causes stars to form, age and die more quickly due to gravitational or tidal effects.

Unlike the familiar neat textbook picture, Milky Way galaxy is surrounded by numerous globular clusters and dwarf galaxies orbiting it at random angles relative to the galactic plane. There are also stellar streams, ie, spread out arcs of stars, orbiting our galaxy at odd angles to the galactic plane that appear to be remnants of earlier encounters, much like meteors that follow comet orbits and that are responsible for our meteor showers. Some of these stellar streams are associated with known satellite systems such as the Sagittarius Dwarf Elliptical Galaxy, which orbits at 90 degrees to the galactic plane. Most of these have lower metalicity than the Milky Way. Andromeda, the nearest spiral galaxy, is moving toward the Milky Way and will merge with it in about 4.5 billion years. There are hints that the Andromeda galaxy extends far beyond the visible edge and that may indicate our own galaxy is much larger than previously suspected. Such collisions and mergers may contribute to star formation, aging and deaths that produce heavier elements through nucleosynthesis.

Is the theory that nucleosynthesis occurs only in stars and supernovae science or philosophy? It is philosophy. Although mathematical models and calculations show that it is possible to form heavier elements inside stars and in supernovae, and other evidence seem to confirm it, is this the origin of all heavier elements, or is stellar nucleosynthesis only one contributor to an already extensive community of elements? There is no way to really know. What if all known elements have existed throughout the history of the universe and their numbers are only increased through stellar nucleosynthesis and supernovae? How would that change our understanding?

ORIGIN OF GALAXIES, STAR SYSTEMS AND PLANETS

The Big Bang theory, according to mathematical modeling based on pure reason, assumes an isotropic and symmetrical universe. That is, a very uniform (isotropic) universe whose laws work the same way forward or backwards (symmetrically), including time. Since entropy[36] is always expected to increase, some cosmologists think that one of the things that differentiates the past from the future is the increase of entropy with time. As such, time apparently loses its symmetry. How is this possible if the theory of symmetry is correct? How can galaxies and larger structures form from an isotropic, very uniform, universe just after the Big Bang that is assumed to have resulted in a very uniform cosmic microwave background?

Cosmologists can only speculate that there must have been early non-uniformities that allowed gravity to organize knots of matter into ever larger structures.[37] So, can this and the assumption of universal isotropy both be true? In the cases of both isotropy and symmetry, exceptions to the basic assumptions of the Big Bang theory are necessary in the real world. Another assumption is that, not just matter and energy are conserved, but that information is also conserved, ie, neither created nor destroyed, only changed in form. The assumption about information is borrowed from the mathematical models, (assumptions), of particle (quantum) physics.

This theory states that information about the physical state of matter at any time should allow us to determine its state at any other time. If information is always conserved and nothing can escape a Black Hole, not even light, how is information still conserved? This is called the Black Hole Information Paradox. Some cosmologists theorize (guess) that black holes emit radiation that, over time, will evaporate them completely, returning the "lost" information. All of this is still hotly debated by theoretical quantum physicists and cosmologists.

So how did galaxies, clusters and larger structures form? The formation of galaxies and larger structures from an extremely uniform early universe into the lumpy, clumpy arrangement we see today is a great mystery and source of debate to believers in the Big Bang theory. The leading theory is that quantum or dark matter fluctuations in the early universe started localized gravitational attraction that grew into all that we see today. There is some obfuscation on the subject of isotropy that assumes the present universe, on a large enough scale, is uniform. While this may be true, it does not answer the question of the assumed extreme isotropy in the early universe. So we have two types of isotropy that are often used as if they are the same. To see uniformity of arrangement today, it is necessary to assume that the anisotropy (non-uniformity) that occurred in the very uniform early universe was also distributed uniformly.

Galaxies are surrounded by spherical halos of molecular hydrogen that is constantly being pulled in by gravity. Deep space X-ray astronomy uses holes in these hydrogen clouds to see distant objects because molecular hydrogen shows up as haze in the X-ray images. These clouds also have metalicity (heavier elements) similar to the galaxy but in lower amounts. This material may continually resupply the galaxy to form new stars along with the gas and dust already within the galaxy. Galaxies collide with other galaxies and may engulf smaller ones. Thus, galaxies may grow by acquisition. Collisions and acquisitions may also trigger bursts of star formation through tidal forces.

LARGE SCALE STRUCTURES

Another supposed manifestation of dark matter is the gathering and movement of galaxies in clusters and walls. Calculated from their relative redshifts, these clusters have been mapped in 3D. Based on this, computer generated maps of dark matter have been made based on their supposed gravitational fields. Let me reemphasize that dark matter does not have to be exotic, only cold dark ordinary matter not illuminated by stars. Intergalactic space may not be as empty as we think. There may be more matter, thinly dispersed between the widely spaced galaxies than within them. In the blackness of space, the only things we can see are stars and those things that are illuminated by them or that block their light.

Since very few stars exist independently of galaxies, and indeed are probably born within them, it is not unreasonable to assume that un-illuminated matter in the vast distances between galaxies may far outweigh all the matter in the galaxies. Remember, that there is an estimated 7×10^{29} (700,000 trillion, trillion) times as much space as matter.[38] Even a tiny amount of widely dispersed matter would add greatly to the total mass. If we assumed 10 times the assumed amount of matter, that would still be an estimated 7×10^{28} (70,000 trillion, trillion) times as much space as matter, neatly accounting for an additional 90 percent of mass without appreciably affecting visibility.

If space is filled with solid objects such as dust and/or gas, wouldn't that attenuate the light from distant galaxies or obscure them entirely? Not necessarily, if the particles are relatively small. Like the lenses of sunglasses, only some dimming may occur without a significant contribution of extra spectral lines. Neutral density filters, composed of fine solid grains in a film, are used in photography to attenuate the light without shifting or adding to its spectrum. Light from distant galaxies is dimmer than would be expected, and this is attributed to recessional motion in the expanding universe model. It could also be explained by attenuating intervening matter.

Cosmic microwave background radiation, (CMB) mentioned earlier, could just be the residual temperature of this unseen matter, rather than the afterglow of the Big Bang. It has a relatively featureless blackbody spectrum,[39] like a heat signature, and is very uniform across the entire sky except where it is blocked or overwhelmed by the Milky Way. The CMB is very uniform but has patterns of variability. Some of these slight variations have been traced to specific galactic clusters.

FIRST CAUSE

Then there is the problem with what occurred before the Big Bang, and what caused the original explosion. Where did the energy originate and what caused this energy to expand violently into a new universe? Here we are faced with the problem of the original necessary cause; the uncaused cause that theologians say is God or the Creator. Since everything in the universe has a cause, what caused the original cause? Without an uncaused cause behind it all, we just have an infinite progression of causes behind causes, which makes no sense and is no explanation at all. It is a great mystery. In order to start a series of causes, something outside the series must start it. This would be something or someone outside of time and space.

Some people who reject the existence of God or gods see the Big Bang theory as a creationist theory. Most cosmologists, who otherwise refuse to allow for unexplained phenomena lest God get a foot in the door, accept that they cannot know what happened at or before the Big Bang. Rather than allow for the possibility of a necessary first cause that hints at God, they steadfastly remain mute or completely abandon causality, on which all of science and logic are founded. Abandonment of causality is seen in the most popular quantum physics interpretations. See next section. Here again, it is philosophy, not science. It is a belief, dare I say a religion.

> In view of such harmony in the cosmos which I, with my limited human mind, am able to recognize, there are yet people who say there is no God. But what makes me really angry is that they quote me for support of such views.
>
> *- Albert Einstein*

Since everything in the universe is expected to progressively change, finely tuned fixed physical values that allow for formation of stars and galaxies and for life to progressively evolve from it are a problem for these theories. The very values, in their thinking, must also change with time. One attempt to obfuscate this point and explain why the universe has such finely tuned fixed parameters, is to assume multiple universes each with a different set of random physical values. By this scheme, ours is one of the lucky ones that at random ended up with just the right balance to allow for formation of long lived stars and galaxies and for life to progressively evolve from it.

> I find it quite improbable that such order came out of chaos. There has to be some organizing principle. God to me is a mystery but is the explanation for the miracle of existence, why there is something instead of nothing.
>
> *- Allan Sandage, cosmologist*

Another attempt to push away any consideration of a first cause is to assume an expansion after a Big Crunch collapse of a previous universe. Neither multiverses nor a previous Big Crunch are detectable and neither of these schemes really answers the ultimate question of first cause or the existence of something rather than nothing.

CONCLUSIONS AND AFTERTHOUGHTS

As you can see, the redshift paradigm of the expanding universe from the beginning in a Big Bang event is fraught with problems, questions and mysteries. What would happen if we assumed a different paradigm or no paradigm restrictions? Could we then have a simpler picture of the universe that needs fewer ad hoc (special case) explanations? If not constrained by a fixed paradigm, could we then learn more about our universe? Needless to say, anything that rejects or opposes the standard model would meet with emotional and political opposition by the status quo cosmologists whose careers have been invested in the present picture. Funding and publication would still be a problem unless the cosmological community reversed its present exclusionary attitudes.

My purpose here is not to tear down science, but to open it up to more exploration and discovery, unconstrained by fixed paradigms. There should be no question that cannot be asked, and no phenomena that cannot be examined freely without prejudice. We wouldn't want to throw out the standard model and start over from scratch, because it may be perfectly fine. However, there should be an open forum for unconstrained exploration, questioning and discussion without repercussions. Scientific and academic freedom is the goal, leading to new discoveries, answers to problems and greater understanding.

PART 6

PARTICLE PHYSICS
AS EVOLUTION FROM NOTHING

CHAPTER 14

QUANTUM MECHANICS, THE HEART OF PARTICLE PHYSICS

Quantum mechanics (QM) or quantum theory, which is based on complex mathematics, tries to describe and explain the odd behavior of particles and forces in the atomic and subatomic realm. Since like charges repel each other and opposite charges attract, it was a great mystery how positively charged protons (and neutral neutrons) can be held together in a compact nucleus without flying apart, and why negatively charged electrons do not crash into the positively charged nucleus, but instead are held in stable orbits around it.

Obviously, to overcome electrostatic attraction and repulsion some other forces or principles are at work that do not exist or manifest[1] in larger size ranges. In this theory, things don't happen in a smooth (analog) manner but in a punctuated (digital) manner and particles exist or move between one allowed state and another based on discrete packets (or quanta)[2] of energy that they absorb, emit or carry. This behavior is manifest in the atomic spectra that consist of forests of absorption and emission lines at specific wavelengths representing allowed transitions of electrons from one orbital (energy level) to another.

In quantum theory, subatomic particles are described as both particles and waves simultaneously. This is referred to as wave-particle duality. To help you picture this, you might think of them as either vibrating particles or particles riding stabilized or standing waves.[3] All types of energy, including subatomic binding forces, are also defined as both particles and waves, so that matter and energy are treated as if they are the same thing. Indeed, electromagnetic energy such as light, heat and X-rays, are all defined as particles (photons) and as waves having definite wavelengths.[4] Both subatomic particles and photons sometimes act like waves and sometimes like particles, depending on how they are tested or detected. Shorter wavelengths (or the higher frequencies) are more energetic. For example, X-rays can disrupt cells and damage molecules but visible light cannot. To explain this it is assumed that the shorter the wavelength, the larger the photon (particle).

Atoms consist of positively charged protons and neutral neutrons in a tightly bound nucleus surrounded by clouds of negatively charged electrons in specific allowed orbitals or energy levels also called electron shells. These can be thought of as either distances or energy levels. Electrons move around the nucleus at high speeds so that their exact location at any one moment is not known precisely without measurement. The likelihood of finding a given electron at a particular place in its orbital is defined by a probability, thus the cloud or shell. Orbitals are only allowed where the orbital path (circumference) is a whole number multiple of the wavelength of the electron, thus setting up constructive interference that stabilizes it. Otherwise, destructive interference would destroy its stability.

Let's use sound waves to illustrate positive and negative interference. Sound waves are similar to ripples in water where crest follows trough follows crest. When the crests and troughs of multiple sound waves exactly coincide they reinforce each other. This is called resonance or positive interference. In the design of stringed instruments for example, ideally the dimensions of the resonance chamber exactly match the range of tones produced or their harmonics (exact multiples or fractions), so that the waves bounce back and forth reinforcing each other to build a fuller sound. When crests and troughs do not match up, the sound is damped by negative interference. Noise canceling headphones work by adding a trough for every crest received and vice versa. Similarly, wavelengths of light or electrons can also set up positive interference, usually called resonance.

An electron jumps from one allowed orbital to another by absorbing energy (a photon) at a specific energy (wavelength). The absorbed photon at a specific energy level is called a quantum, thus quantum theory. The electron will also fall from this excited state back to its more stable ground state orbital by emitting a quantum of energy. Because the electrons in any one atom can only absorb discrete energy quanta, a spectrum of light interacting with an atom shows absorption and emission lines at specific wavelengths that can be used to identify the element.

That is all well and good, but it gets weirder. In the widely accepted Copenhagen interpretation of quantum mechanics, a particle is said to not have a fixed state but exist in a smeared out multiplicity of states at once until a measurement is taken, when it collapses into one state. This is the principle of superposition. Because an electron can be found in any of the probability allowed shell locations, this interpretation assumes that the electron really is at all the locations at once and only assumes a fixed state when measured. This assumption extends to the other characteristics of the electron such as spin or momentum, as well as to whole atoms or other subatomic particles.

The Copenhagen interpretation of quantum theory also says the electron exists in a multiplicity of superposed states at one or the other orbital level but does not exist anywhere between. When a quantum of energy is absorbed the electron is said to pop out of existence in the original shell and simultaneously pop into existence in the new shell. But, since the electron shell defines a probability, and most of the time the electron exists in one of these shells, the probability of finding it anywhere between is statistically infinitesimal. It is said not to exist there, and it is thus called "forbidden." Is it only an extremely small probability or are we talking about its actual existence? The Copenhagen interpretation of quantum theory says it is the latter. Other interpretations of quantum theory differ as to what actually happens. See other interpretations below.

This is only a very simplified peek at the strange quantum world. Let's back up and look at the history and proposals that led to today's quantum theory. Max Planck calculated that the energy and the frequency of light were related by a constant, now called Planck's constant, (h), with a value of 6.625×10^{-34} joule-seconds. This means that for any frequency of light, its energy, W, can be determined by multiplying the frequency by the constant. ($W = h\nu$) where W is the energy, h is the constant and ν is the frequency.[5] It turned out later that this same constant could be used in determining other atomic and subatomic parameters, and is thus fundamental to calculations in quantum mechanics.

Einstein theorized that light was not just a standing wave, but also a particle based on the photoelectric effect in which light of a high enough energy, when shown on a metal, kicked off electrons. Thus began the idea of wave-particle duality, later extended to the electron by Niels Bohr and then to all subatomic particles by Louis de Broglie.

Based on experiments where most alpha particles[6] shot at a gold target went straight through but a few bounced back, Ernest Rutherford first proposed an atom similar to our solar system in which most of the mass is contained in a compact nucleus surrounded by orbiting electrons. In an attempt to explain the spectrum of atoms Niels Bohr proposed the present picture of the atom as a compact nucleus with elections orbiting that could only absorb specific wavelengths of light and therefore could only occupy certain fixed energy levels, ie, orbitals. In trying to measure these discrete orbitals and their electron locations and momenta,[7] it became apparent that measurement of any kind disturbed the system so that only one of two coupled parameters could be determined at any one time, eg, position and momentum (or speed).

This led to the Heisenberg uncertainty principle, which states that it is impossible to know both the position and the momentum of any one subatomic particle at the same time. The system is disturbed by measurement because measuring subatomic particle parameters is like administering eye drops with a fire hose. Because the subatomic particles are so small compared to any means of measuring their parameters, what is measured is in a disturbed condition. The Heisenberg uncertainty principle was meant to be a statement of experimental limitations, not that location and momentum (or other coupled parameters) did not exist in a fixed state at the same time. However, Bohr, and others interpreted it that way, assuming that atoms or atomic particles were never in a fixed state until measured, and that uncertainty is a fundamental characteristic of subatomic particles, not just an experimental limitation. Thus they have substituted ontology (being) for epistemology (ability to know). Heisenberg never accepted the principle of superposition or non-locality claimed in the Copenhagen interpretation.

Edwin Schrodinger provided the mathematical equations[8] for the behavior of electromagnetic waves that are used in quantum mechanics to account for their amplitude (crest height) as well as the frequency. These partial differential wave equations are linear, that is they can be plotted as straight lines on a graph. Superposition is a concept in mathematics stating that in linear equations all of the contributing factors must add up to the net effect of each factor individually.

Since Schrodinger's equations for waves are linear, it is assumed that their application to subatomic particles is also linear. From there it is a leap of faith to assume that particles don't just have the capability of being in different states, but that they are simultaneously in all possible states at once. Instead of just being a mathematical concept, superposition now was applied directly to subatomic particles in a real physical sense.

However, Schrodinger did not agree with this Copenhagen interpretation of quantum mechanics. He came up with an example within everyone's sphere of experience that illustrated the absurdity of their assumed superposition. This was the famous Schrodinger's cat thought experiment. He set up the experiment so that a cat in a closed box could be either alive or dead, depending on whether a radioactive particle spontaneously decayed setting off a mechanism that released a deadly poison gas. In this thought experiment, using the Copenhagen interpretation of superposition, since we don't know what state the cat is in until the box is opened, the cat is both dead and alive until it is opened at which time the cat becomes either dead or alive. The act of observing somehow must cause the cat to assume either a dead or alive state. In all other realms, this would be called magical thinking. Meant to point out the weakness or absurdity of superposition, it has been used to illustrate the opposite through convoluted "reasoning" to make it fit the Copenhagen or similar interpretations.

The idea of instantaneous communication and action at a distance is a consequence of this assumed superposition where particles did not assume a fixed state until observed. By Pauli's exclusion principle, no two electrons in the same orbital can be in the same quantum state. Each must differ in some way, for example they must have opposite spins. The two particles are said to be entangled since each must be in the opposite state to the other. If one of the electrons is emitted and travels relatively far away, when one of the electrons is measured (observed), it collapses into a fixed state and simultaneously the other one collapses into the opposite state that can be confirmed when it is measured. This implies speed of communication faster than the speed of light, the assumed upper limit of speed.[9]

Einstein and others never accepted the Copenhagen interpretation of superposition, and called this "spooky action at a distance."[10] He had dealt with action at a distance of gravitational attraction in his theory of relativity. He solved the gravity problem by theorizing warping of space-time by mass. In that way, the attraction of gravity is caused by the gravity wells created by mass. In a paper,[11] Einstein and colleagues described a thought experiment that presented a paradox illustrating that the Copenhagen interpretation was either wrong or incomplete. The bottom line to their argument was that any theory must completely explain the phenomena and must also correspond to reality or it can't be a complete theory.

Einstein thought that quantum action at a distance was an illusion based on the assumption of superposition, aka non-locality. If particles are assumed to have fixed states, although unknown to an observer, the action at a distance is no mystery. It only implies that entangled states, eg, opposite spins, persist after separation. When one of the particles is measured, you automatically know the state of the other since they must be the opposite of each other, whether together of separated. Einstein spent the latter part of his career trying to prove this.

We have discussed the Copenhagen interpretation of quantum theory, but there are other interpretations we should look at. There are at least a couple of dozen other interpretations. The most popular, among a long list, (see table following), are the Copenhagen interpretation and its variants, the many worlds interpretation and the ensemble interpretation. Variants of the Copenhagen interpretation involve either the observer or the cat (as observer and participant) as being parts of the quantum system. Another, the many worlds interpretation, is even more speculative. In this scenario, each time a subatomic particle collapses and chooses a fixed state, reality splits in two and both possible realities still exist, but in different undetectable dimensions. Think of this as a time series of pictures or a strip of movie film. At the decision point, the one series becomes two, and at the next decision point, becomes four, et cetera, ad infinitum.

The ensemble interpretation states that quantum mechanics can only be applied to statistically significant numbers of particles, not to individual particles. Since the wave equations describe probabilities, it would be meaningless to apply them to single particles. This is the interpretation favored by Einstein but is discounted by leading QM physicists. Similar realistic interpretations such as those proposed by de Broglie-Bohm and science philosopher Karl Popper assume real particles with real positions and real wave functions that do not need to collapse upon measurement. I tend to prefer these theories because of their realism.

> The attempt to conceive the quantum -theoretical description as the complete description of the individual systems leads to unnatural theoretical interpretations, which become immediately unnecessary if one accepts the interpretation that the description refers to ensembles of systems and not to individual systems.
>
> - Albert Einstein, *Albert Einstein: Philosopher-Scientist*

As a consequence of the probabilistic view of the subatomic world, quantum theory leads to a conclusion that events are not deterministic, but rather are indeterminate; that they just happen without actual connections between cause and effect. If deterministic, events in the past must predict future events as causal antecedents. In the macro or real world, everything has a cause or causes. Determinism is the accepted view or apparent state of the real universe because, knowing the mass, position and the momentum of a (larger) body, plus all of the influences on it and the mathematical equations governing its movement, one can (in theory) calculate its position and speed at any other time in the future or the past. This is the basis of celestial mechanics by which planets and other bodies are tracked.

However, this is over-simplistic. It works well for two bodies acting on each other, whether gravitationally for planets or electronically for subatomic particles. When more than two bodies in motion are considered, there is a degree of uncertainty introduced by their complex interactions that is difficult or impossible to eliminate. This is called the Three Body Problem. Projecting outcomes far into the future or past becomes increasingly uncertain the farther you go. The Three Body Problem and the use of partial differentials introduce uncertainties in mathematical calculations concerning subatomic particles.

For individual particles or atoms with more than one electron in the outermost orbit being tested, the calculations become maddeningly complex. Most of the work in this area is done on either photons (single units), hydrogen (with one electron) or hydrogenic atoms with one outer orbital electron. Quantum mechanical calculations using partial differential equations and wave functions involving things like Hamiltonian operators, Eigen functions or Hilbert space are notoriously difficult to solve even for two bodies. Introducing more than two bodies, they become almost impossible without uncertainty and errors. The question is: since we don't know for sure what the outcome according to QM will be, is it really indeterminate or are there certain things we don't or can't know about the system that only makes it LOOK indeterminate? If it were possible to know all of the parameters and influences without disturbing the system could we, with certainty, predict outcomes?

MAJOR QUANTUM MECHANICAL INTERPRETATIONS OF MAINSTREAM PHYSICS

THEORIES	REAL PARTICLE/ WAVE	BRIEF DESCRIPTION OF MAIN FEATURES
Copenhagen	No	Subatomic particles/waves exist in a smeared out superposition of all possible states until measured/observed
Many Worlds	No	Each time a measurement is made the superposition of states collapses into all possible states, each in its own dimension, all but one of which are unobservable.
Ensemble or Statistical	Yes	Mathematical probabilities can only be applied to statistically significant ensembles of systems, not to individual systems or particles.
DeBroglie-Bohm Theory	Yes	Particles have definite positions guided by the real wave function, which never collapses
Consistent Histories	No	QM predicts the probability of each alternative history.
Relational	No	Different observers see different states which describe not the state but the relationship to the observer
Elementary Cycles	Yes	Particles have recurrences in space-time so that QM is the statistical description of each of these cycles
Transactional	Yes	The standing wave is the result of the past and future waves but does not depend on the observer.

Stochastic Mechanics	Yes	Classical derivation and interpretation of the Schrodinger equations
Objective Collapse	No	Collapse of the wave function occurs randomly and does not involve the observer
Von Neumann/ Wigner	No	Consciousness causes collapse into a fixed state
Many Minds	No	Like many worlds except that it involves consciousness
Quantum Logic	No	A type of logic that tries to reconcile Boolean logic with QM
Quantum Information Theories	No	Variant 1. Ontological: information is existence; Variant 2. Epistemic: QM describes the observer's knowledge, not existence.
Modal	Yes	Based on Schrodinger wave equations but lacks collapse; particles have definite but dynamical states at all instants; measurement only reveals the state at that time.
Time-Symmetric	No	QM modified to include symmetric time so that past and future are equivalent and all operations are reversible in time.
Branching Space-time	No	Like many worlds but splitting of space-time, not just splitting of wave functions.
Popper	Yes	Non-locality defies common sense and the known laws of physics. Determinism and real particles and waves with real values.

In addition to these popular interpretations, there are a dozen or so minority interpretations and several alternative theories not recognized by main stream particle physics. Minority interpretations include the following:[12] Calogero conjecture, Semiotic, Prowave, Pondicherry, Quantum Mysticism, London or Ticker Tape, Incomplete Measurements, Montevideo, Quantum Bayesianism, Synchronized Chaos, Vaxjo, Dimensional. In addition to these, there are alternative interpretations not recognized by most of the particle physics community. Can a theory with so many varied interpretations be considered valid or complete?

So, is quantum theory or quantum mechanics science or philosophy? Well, it is a little of both. Many of the basic concepts on which it is based such as Bohr's model of the atom, Planck's constant, Schrodinger's wave equations, Heisenberg's uncertainty principle and Pauli's exclusion principle, are well established by experiment. However, many of the interpretations seem to confuse mathematical equations with reality, and in some cases ontology with epistemology (being with knowing). Remember, mathematical equations can only describe reality, not create it. They are not reality in themselves, but are attempts to describe reality. This is a classical category mistake whereby the description and the thing are confused. In this case, some interpretations confuse the mathematical theory for reality itself.

Being possible mathematically does not necessarily make it probable or even possible in the real world. In this way of thinking, if mathematical operations are reversible then reversal of time itself must be possible; if equations can work in multiple extra, non-detectable dimensions, then these dimensions must really

Mathematicians deal with possible worlds, with an infinite number of logically consistent systems. Observers explore the one particular world we inhabit. Between the two stands the theorist. He studies possible worlds but only those which are compatible with the information furnished by the observers. In other words, theory attempts to segregate the minimum number of possible worlds which must include the actual world we inhabit. Then the observer, with new factual information, attempts to reduce the list still further. And so it goes, observation and theory advancing together toward a common goal of science, knowledge of the structure and behavior of the physical universe.

- Edwin Hubble, "The Problem of the Expanding Universe"

exist. Only when mathematics faithfully describes reality can it be considered validly scientific. Mathematical descriptions that deviate from reality must either be wrong, misinterpreted or incomplete descriptions of reality. This was the point that Einstein and others argued.

I hope the reader will appreciate that I have struggled to describe the basics of quantum mechanics while eliminating most of the complex mathematics on which it is based. Although it is unnecessary to illustrate my point, I could delve further into this subject in such areas as the forces holding subatomic particles being described as exchanges of virtual particles, properties of quarks which comprise protons, neutrons and mesons, the assumption that vacuum energy[13] creates virtual particles in particle-antiparticle pairs, Higgs boson or Higgs field that is assumed to give all other particles their mass (because QM calculations assume all particles are massless and dimensionless). This would serve no purpose other than to confuse the reader with superfluous complexity intended not to instruct but to impress. I think this abbreviated discussion of quantum mechanics is sufficient to illustrate that the theories of particle physics are far from settled or proven. All of these other parts of particle physics are based on QM and generously use the same interpretations, usually Copenhagen.

The field of particle physics is filled with many unsolved mysteries just waiting for open minded scientists to come along and unlock them. Appeals to the best explanation or absence of other explanations are only a form of appeal to ignorance fallacies and can readily be made obsolete by new insights. Remember, the theory of epicycles of pre-Copernican astronomy was considered the only or best explanation at the time. This goes for appeal to authority as well.

Enshrinement of famous personalities as infallible can interfere with the search for the truth. By concentrating on unrealistic or questionable interpretations of subatomic reality, many possibilities for exploration are excluded or at least de-emphasized. By questioning or side-stepping some of the orthodoxy, new avenues of research should be possible. The fact that there are other interpretations still being debated is a healthy sign, although most research grants are given to support the status quo Copenhagen interpretation. Because there are sufficient numbers of physicists questioning status quo interpretations, we may be on the threshold of opening a whole new world of exploration leading to a more complete knowledge that is based on a search for the truth, not status quo politics.

PART 7

A MODERN EXAMPLE

CHAPTER 15

CLIMATE CHANGE: PHILOSOPHY DISGUISED AS SCIENCE

For a current example of philosophy and/or politics disguised as science, we need look no further than the climate change debate. Regardless of the merits of the case for the anthropogenic global warming theory (AGW) that manmade carbon dioxide is the cause of climate change, the way it has been advocated is more akin to a political campaign than to a dispassionate search for truth. Political action is advocated that would drastically change our world, crippling industry and technological progress while leaving developing nations to flounder in their poverty. The two- pronged approach of this philosophy is to curtail both technology and population.

Developed countries are said to hog all the resources at the expense of developing countries. People are said to be the problem, and advanced societies must be brought down to near subsistence levels while primitive societies are not raised from their squalor. But is any of this true? Is it science or is it politics? Unfortunately, it is more about politics, philosophy and belief than about science. Are there too many people, and are subsistent societies cleaner and less ecologically harmful? Are developed nations really hogging all of the resources at the expense of developing ones? The answer to each of these questions is NO.

TYPICAL POLITICAL TACTICS EMPLOYED INCLUDE:

1. Appeal to authority, (a logical fallacy): the "consensus of scientists" with only a very small, elite group deemed qualified to understand or comment on it.
2. Appeal to ignorance (a logical fallacy): It must be increasing CO_2 (Carbon Dioxide) because we can't find any other cause – but we aren't looking very hard at things like wind and water cycles or solar activity.

3. Depend on statistics and computer models instead of real historical facts and experimental data. Remember GIGO – Garbage In, Garbage Out. A model is only as good as the data used or left out, the type of mathematical calculations based on that data and assumptions and conclusions made.

4. Use fear to sell the agenda: Use warnings of catastrophic consequences if action is not taken immediately by governments, industry and individuals based on models of a poorly understood atmospheric and planetary system. (The con: "special today just for you but you must call within the next 30 minutes or you'll miss out." Or "you can save the planet, so call your congressman today before it's too late.")

5. Use guilt and shame to get people, governments and industries to "go green" and curtail the activities that use fossil fuels or otherwise emit CO_2.

    a. Do you use incandescent light bulbs? Then you're killing the planet because you're consuming power from fossil fuel driven power plants. The government must phase out incandescent light bulbs in favor of LED or compact fluorescents, (which contain mercury, a primary pollutant); we must regulate power plants to insure maximum efficiency regardless of increased cost to the consumer, which hurts the poor most.

    b. Do you eat beef? You're killing the planet because of methane from cows. The government must regulate the methane from cows.

    c. Do you fly, drive or use a ferry or train? You're killing the planet because of fossil fuel consumption. The government must demand more efficient transportation – even if CAFE[1] standards demand lighter and less safe vehicles that are killing people.

    d. Do you use manmade fibers or plastics in any form? You're killing the planet because it takes fossil fuels to produce them – never mind that most of these things get put in a landfill, which is a form of sequestering carbon. The government must regulate the industries that produce them – and the landfills, too.

    e. Do you use paper products? You're killing the planet because trees that could consume CO_2 are cut down to produce paper. Never mind that trees are farmed and harvested and new trees are planted to more than replace those used. Younger trees consume CO_2 at a faster rate per ton than older trees.

6. Use the press to promote their views and disparage the opposition in the form of propaganda. (TV, radio, internet, movies, books, magazine and journal articles, newspapers)

   a. Present a parade of "experts" and dire predictions as absolute settled facts, not as projections of a computer model.

   b. Sensationalize and exaggerate any "fact" that supports the global warming theme and downplay or fail to report on things that don't.

   c. Declare that the polar bears are drowning because the sea ice is melting. Never mind that polar bears can swim up to 60 miles between feeding areas, that there is no net loss of sea ice over time and that polar bear numbers are increasing.

   d. Have a storm, flood or drought? Blame it on climate change. Make it sound biblical in proportions and the worst in history.

   e. Have a problem with mosquitoes because of a particularly wet spring? It must be global warming.

   f. Are the seas rising at the same rate they have been for centuries?[2] Oh, my God, our cities will soon be underwater and we're all going to drown!

7. Demonize those who disagree as "deniers" with the unspoken implication that they are on a moral level with holocaust deniers. Climate experts that aren't on board with the whole global warming scenario and who have DATA to back it up are called "just weathermen" who are unqualified to comment, even though many of them have better credentials than many of the AGW proponents.

8. Exclude from publication or grants, any research that doesn't agree with their conclusions, and then declare that there are few peer reviewed papers on the other side. Never mind that government funding is overwhelmingly on one side. Journals such as *Science* and *Nature* have become advocates instead of unbiased scientific publications.

9. Hide raw and analyzed data and analysis methods from other researchers who wish to verify the work. Real science always shares data and methods with other researchers so their results can be verified. This one includes the "massaging" of the data to say something it doesn't. The Climategate scandal was all about hiding the data and massaging it to eliminate the Medieval Warm Period and the Little Ice Age and to create a "hockey stick" that was used to alarm governments into drastic control measures. When other researchers

finally got hold of the (massaged) data[3] and analysis methods, it was discovered that any random set of numbers, when plugged into the formula, produced a similar "hockey stick." This showed that the analysis algorithm on which the computer models were based was worse than worthless.

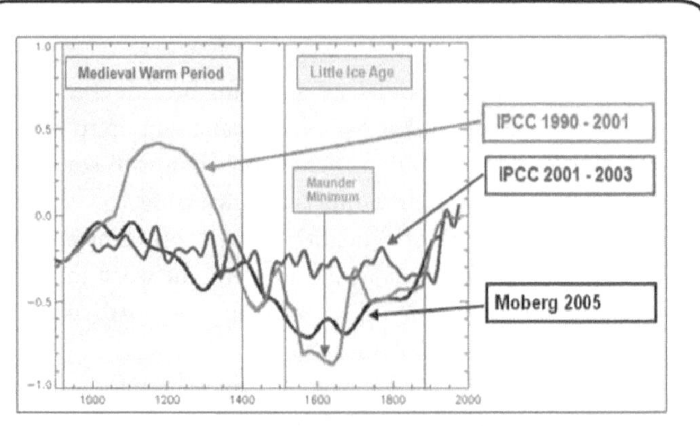

Comparison of graphs published by earlier and later versions of the IPCC Assessment on Climate Change. Note that the later graph eliminated both the Medieval Warm Period and the Little Ice Age.[4]

Note that the vertical temperature scale is less than a degree above and below today's temperature, which is set near zero on the graph.

The data for the line marked Moberg is available through NOAA website and is from "Highly variable Northern Hemisphere temperatures reconstructed from low and high-resolution proxy data" *Nature,* Vol. 433, No. 7026, pp. 613 - 617, 10 February 2005. Anders Moberg[1], Dmitry M. Sonechkin[2], Karin Holmgren[3], Nina M. Datsenko[2] & Wibjörn Karlén[3]

1 Department of Meteorology, Stockholm University, SE-106 91 Stockholm, Sweden

2 Dynamical-Stochastical Laboratory, Hydrometeorological Research Centre of Russia, Bolshoy Predtechensky Lane 11/13, Moscow 123 242, Russia

3 Department of Physical Geography and Quaternary Geology, Stockholm University, SE-106 91 Stockholm, Sweden

10. Ignore other factors that may contribute to or mediate the supposed effects of manmade carbon dioxide (such as water vapor, high and low altitude clouds, methane, increased plant growth, ocean sequestering or release, solar activity cycles, and precession of Earth's tilt).

Manmade global warming, or climate change as it has come to be called,[5] is said to be an established fact, and the consequences are dire unless global governments act now to mitigate its effects. The polar ice caps and glaciers will melt away; the oceans will rise and drown coastal and island regions; droughts, floods, storms and temperatures will all increase and millions, dare I say billions, will die. But is any of it true, and is it science? The answer is NO. While there has been general warming since the Little Ice Age of the seventeenth and eighteenth centuries overlain with periods of lesser heating and cooling, is the change good or bad? Is it unusually rapid now and will these alleged trends continue into the future? Is it extreme enough to cause the dire effects predicted and should global governments act now to prevent the predicted disastrous consequences? It is important to know if these modeled projections are reliable predictions and if real, is real science involved in any meaningful way. What do we really know about it?

WHAT IS CLAIMED:

1. Global warming and/or climate change are established facts.
2. Manmade carbon dioxide (CO_2) is the main cause of global warming.
3. Carbon dioxide is important because it has a forcing effect on other factors such as water vapor.
4. Manmade CO_2 levels have been rising rapidly due to increased industrialization and populations since the 1950s.
5. Developed countries are to blame for increased manmade carbon dioxide because they use most of the fossil fuels.
6. Temperatures are hotter now than they have been in the last 100,000 years.
7. Temperatures have risen faster in recent years than ever before.
8. The world is in danger of catastrophic consequences such as sea level rise, growing deserts, worse storms, droughts and floods.
9. The oceans are becoming more acidic due to the increased CO_2, so that corals and other animals are being harmed or killed.

10. World governments must take drastic action now to prevent further warming.
11. There is a consensus among climate scientists on the causes and dire consequences of global warming.

EXAMINING THE CLAIMS

Let's look at each of these claims one at a time. Many of the claims are only partially true or are false. In this section I have included many charts of real data to show the truth and to refute the claim by warmists (proponents of AGW) that skeptics have no data. Many of the charts are from a review article, "Environmental Effects of Increased Atmospheric Carbon Dioxide," circulated by the Petition Project. All of the data in this report is from peer reviewed scientific literature cited at the end of the article in 132 references.[6]

Claim 1: Global warming and/or climate change are established facts.

Truth: This is mostly true but implications are false. Climate is always changing. The world has been recovering from the Little Ice Age since the eighteenth century. There was a period of cooling from the 1940s through the 1970s, leading modelers to predict a coming ice age. For example, James Hansen of NASA's Goddard Institute for Space Studies (GISS) was responsible for modeling and alarming Congress and others about global cooling in the 1970s and global warming beginning in 1988 to the present.

WHO IS JAMES HANSEN ?

He was the head of NASA's Goddard Institute for Space Studies, Earth Sciences Division, 1981-2013. He holds a BA in mathematics and physics, a MS in astronomy and a PhD in physics, not climatology. His major work before becoming the climate change scare monger was in modeling the atmosphere of Venus and the effect of its atmospheric gases and aerosols on warming. The atmosphere of Venus is very different from Earth's atmosphere, being composed largely of carbon dioxide with sulfuric acid clouds. It also does not have the magnetic field that protects earth from much of the solar wind's energy and cosmic rays.

He assumed that aerosols in the atmosphere of Venus trap heat to produce the very high temperatures through a run-away greenhouse effect without considering the solar wind and cosmic ray contributions. He has used the models for Venus to model our own atmosphere by making certain assumptions. He has also massaged United States climate data, which showed no net warming in the twentieth century, to show warming where the original data did not. He erased the 1930s as the warmest period of the century, with 1934 as the warmest of all. He "adjusted" the data to show continued warming in line with the previously "adjusted" global temperature record.

See Source: http://stevengoddard.wordpress.com/hansen-the-climate-chiropractor/

Recent increases have been attributed to Carbon Dioxide (CO_2) produced by humans in the form of both increased burning of fossil fuels and increased populations. There are actually many complex reasons for the recent trend of increased temperatures that are not fully understood. The role of carbon dioxide has been exaggerated and the role of other factors such as solar activity and clouds have been dismissed, de-emphasized or otherwise diminished. Presently, (2015) there has been no warming since 1998 and a possible cooling since 2005.

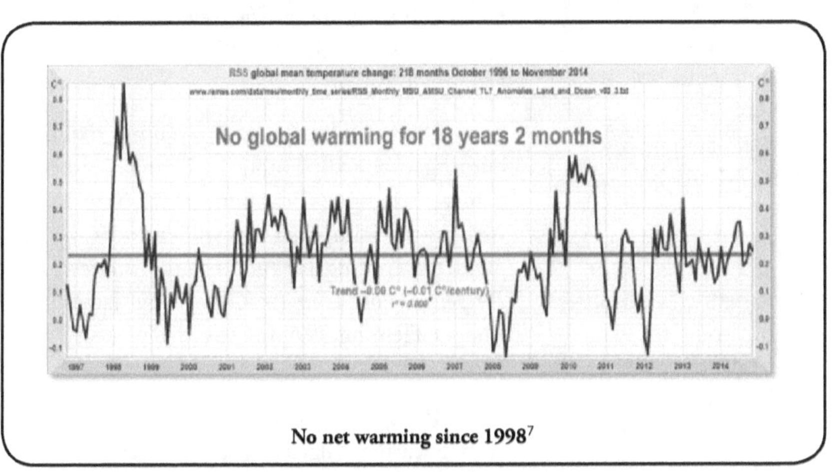

No net warming since 1998[7]

"Since 1895, the media has alternated between global cooling and warming scares during four separate and sometimes overlapping time periods. From 1895 until the 1930s the media peddled a coming ice age. From the late 1920s until the 1960s they warned of global warming. From the 1950s until the 1970s they warned us again of a coming ice age. This makes modern global warming the Fourth Estate's fourth attempt to promote opposing climate change fears during the last 100 years."

- Senator James Inhofe, Monday, September 25, 2006

Here is a more complete timeline straight from the headlines and texts of leading newspapers and other reliable sources, thanks to http://butnowyouknow. net/those-who-fail-to-learn-from-history/climate-change-timeline/ and other reliable documentation as noted below.

- 1872 John Tyndall measured the heat absorption of various atmospheric gases over the entire wavelength range of his heat source. He found that water vapor and CO_2 absorbed more strongly than other atmospheric gases such as oxygen and nitrogen. Oxygen and nitrogen, major components of the atmosphere, had little or no absorption of heat in the range tested. It is important to note that his experiments did not separate the heat into specific wavelengths. See Claim 2 and its chart.

"... if, as the above experiments indicated, the chief influence be exercised by the aqueous vapour, every variation of this constituent must produce a change of climate. Similar remarks would apply to the carbonic acid [CO_2] diffused through the air... they constitute true causes, the extent alone of the operation remaining doubtful."

John Tyndall, Contributions to Molecular Physics in the Domain of Radiant Heat, 1872, Cambridge University Press

- 1895, February, *The New York Times*: "Geologists Think the World May Be Frozen Up Again"

- 1899, Nils Eckholm claims that burning coal will double CO_2 and cause climate change. Eckholm and Svante Arrhenius claim that it will prevent a predicted coming Ice Age. From *Historical Perspectives on Climate Change* by James Rodger Fleming, 1998, Oxford University Press.

- 1902, *Los Angeles Times*: "Disappearing Glaciers... persistency that means their final annihilation..."

- 1912, October, *The New York Times*: "Prof. Schmidt Warns Us of an Encroaching Ice Age"

- 1923, *Chicago Sun-Times*: "Scientist says Arctic ice will wipe out Canada"

- 1923, *The Washington Post*: "The discoveries of changes in the sun's heat and southward advance of glaciers... possible advent of a new ice age."

- 1924, September, *The New York Times*: "MacMillan Reports Signs of New Ice Age"

- 1929, *Los Angeles Times*: "Is another ice age coming?" "Most geologists think the world is growing warmer, and that it will continue to get warmer."

- 1932, *The Atlantic* magazine, "This Cold, Cold World"

- 1933, March, *The New York Times*, "America in Longest Warm Spell Since 1776; Temperature Line Records a 25-Year Rise."

- 1933, *National Weather Bureau Monthly Weather Review*: "... widespread and persistent tendency toward warmer weather... Is our climate changing?"

- 1938, *Royal Meteorological Society Quarterly Journal*: (Global warming, caused by man heating the planet with carbon dioxide) "is likely to prove beneficial to mankind..."

- 1938, *Chicago Tribune*, "Experts puzzle over 20 year mercury rise... mysterious trend toward warmer climate in the last two decades."

- 1939, *The Washington Post*: "... weather men have no doubt that the world at least for the time being is growing warmer."

- 1952, August, *The New York Times*: "... the world has been getting warmer in the last half century."

- 1954, *U.S. News and World Report*: "... winters are getting milder, summers drier. Glaciers are receding, deserts growing."

- 1954. *Fortune* magazine: "Climate – the Heat May Be Off"

- 1955, Gilbert Plass predicts 3.6° C (6.8° F) warming if CO_2 is doubled.

- 1956, October 28, *The New York Times*: "Warmer Climate on Earth May Be Due To More Carbon Dioxide in the Air," by Waldemar Kaempffert in *The New York Times* "Science in Review"

> "... average surface temperature of the earth increases 3.6° C if the CO_2 concentration in the atmosphere is doubled..."
>
> "The extra CO_2, released into the atmosphere by industrial processes and other human activities may have caused the temperature rise during the present century. In contrast with other theories of climate, the CO_2 theory predicts that this warming trend will continue, at least for several centuries."
>
> - *Gilbert Plass*[8]

- 1959, *The New York Times*: "Arctic Findings in Particular Support Theory of Rising Global Temperatures"

- 1969, February, *The New York Times*: "... the Arctic pack ice is thinning and that the ocean at the North Pole may become open sea within a decade or two."

- 1970, *The Washington Post*: "... get a good grip on your long johns, cold weather haters – the worst may be yet to come... there's no relief in sight."

- 1974, *Time* magazine: "Global cooling for the past forty years"

- 1974, *The Washington Post*: "... weather aberrations they are studying may be the harbinger of another ice age."

- 1974, *Fortune* magazine: "As for the present cooling trend a number of leading climatologists have concluded that it is very bad news indeed."

- 1974, *The New York Times*: "... [global cooling] the facts of the present climate change are such that the most optimistic experts would assign near certainty to major crop failure ... mass deaths by starvation, and probably anarchy and violence."

- 1975, *The New York Times*: "Scientists Ponder Why World's Climate is Changing; A Major Cooling Widely Considered to Be Inevitable"

- 1975, Nigel Calder, editor of New Scientist in *International Wildlife Magazine*: "The threat of a new ice age must now stand alongside nuclear war as a likely source of wholesale death and misery for mankind"

- 1976, *U.S. News and World Report*: "Even US farms may be hit by cooling trend"

- 1981, *The New York Times*: (Global Warming) "... of an almost unprecedented magnitude"

- 1988, James Hansen, Goddard Institute for Space Studies, testifies before Congress that global warming is a fact and that consequences of doing nothing will be dire. IPCC, Intergovernmental Panel on Climate Change, was established by the United Nations in that year with the mission to find a connection between human activity and climate change.

- After that, the media blitz of articles supporting the belief in global warming or climate change are too numerous to list in detail here.

Claim 2. Manmade carbon dioxide (CO_2) is the main cause of global warming

Truth:

a. Carbon dioxide is a minor player in any further warming. It is uniformly distributed in the atmosphere but only absorbs infrared (heat) in a very narrow wavelength range. The CO_2 wavelength range is outside the range of most of the solar radiance that penetrates our atmosphere. It falls roughly inside the wavelength range of temperatures re-radiated when solar radiation heats the Earth's surface. The atmospheric CO_2 already absorbs almost all of the radiation that it can in that range. Most of the warming effect of CO_2 has already occurred in the past and is one of the reasons our planet is not a frozen wasteland. Any increase in CO_2 will have a very minor effect. With CO_2 absorption near saturation, almost all of the re-radiated heat in that wavelength range is already being trapped, so it can have little or no effect on future increases in temperature or supposed forcing of water vapor. (See next claim) With CO_2 essentially eliminated as a source, any increase in temperature must be from some other source.

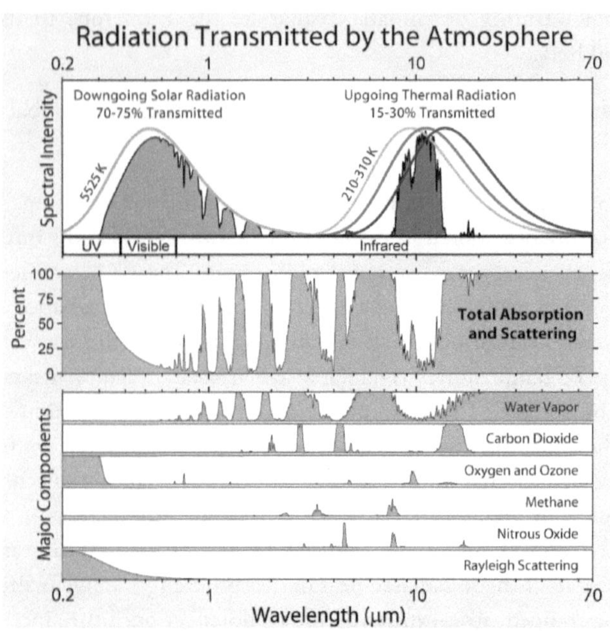

Atmospheric Transmission CC BY-SA 3.0

Source: Robert A. Rohde (Dragons flight at English Wikipedia) - This figure was created by Robert A. Rohde from published data and is part of the Global Warming Art project.[9]

http://www.atmo.arizona.edu/stud ...

This figure requires a bit of explaining. The top spectrum shows the wavelengths at which the atmosphere transmits light and heat as well as the blackbody idealized curves for no absorption. It is a little misleading because the data is not based on actual solar and earth data. It is based on two experimental heat sources, one centered at 5525 K (5252° C or 9485° F), the approximate temperature of solar radiation, and one centered in the range of 210 to 310 K (-63° C to 36.8° C or -82°F to 98° F), the approximate temperature range of re-radiated heat from the earth. In reality, solar radiation power, (Watts/m²/micron), is six million times as strong as the power of re-radiated heat from the Earth.

The other spectra are absorption[10] spectra. The top one shows the relative percent absorption by total atmospheric gases at various wavelengths, (note that this spectrum is practically the inverse of the transmission spectrum above it), and the spectra below that show the absorption wavelength ranges of individual atmospheric gases, not the relative strength of that absorption in reality. As experimental, not real atmospheric, data they can only tell us the wavelength ranges of the absorption, not their relative strengths.

Note that CO_2 absorbs in the 15 micron range,[11] which is within both the range of re-radiated heat and the strong absorption by water vapor, of which the CO_2 peak forms a mere shoulder. This is used to claim forcing of water vapor by CO_2, without regard to the near-saturation absorption level of CO_2. Lesser CO_2 peaks in the 2.7 and 4.3 micron ranges also only contribute in a minor way. The first is completely covered by a water vapor absorption peak and the second forms a shoulder in another water vapor peak. These minor peaks occur in a region where both solar radiation and re-radiation are minimized. Methane and nitrous oxide are also shown to be minor players, having narrow absorption ranges and low concentrations. Note too that ozone blocks most of the ultraviolet light from the sun.

b. Water is by far the most important greenhouse gas/liquid in the form of vapor, high and low altitude clouds, rain and snow, which both absorb and reflect sunlight and re-radiated heat from the surface. Water vapor is not uniformly distributed in the atmosphere, being concentrated near the earth, but strongly absorbs heat in a wide range of wavelengths. More heat means more water vapor evaporating from the oceans. Sounds pretty scary, doesn't it? Contrary to what is assumed by climate modelers, who use this to claim forcing by CO_2, the extra vapor doesn't remain as vapor. It quickly forms low altitude clouds that strongly reflect in-coming sunlight and heat into space. Any re-radiated heat from the surface that may be trapped by clouds is a small fraction compared to the in-coming solar radiation, so blocking solar radiance has a net cooling effect that overwhelms any increases in trapped re-radiation. High altitude clouds tend to trap heat from being re-radiated into space, but have little effect because the increases in cloud cover due to warming are mostly in low altitude clouds.

c. Methane, like CO_2, only absorbs heat in narrow wavelength ranges far from most of solar heat radiance, so that water, with its broad absorbance spectrum, trumps all other greenhouse gases. Like CO_2, methane is at or near its absorbance saturation point so increases would have little effect. While it is true that continued warming could result in release of methane from melting permafrost, it would have relatively minor effect on global temperatures. Methane is derived mostly from decaying organic material and from natural seeps on the land and under the sea, as well as termites and ruminant flatulence. Methane absorbs 29 times as much heat per volume as carbon dioxide but at 1.8 ppbv,[12] (.00000018 percent), compared to CO_2 at 380 ppmv,[13] (0.038 percent), it is recognized as a minor player in greenhouse warming along with Ozone (O_3) and Nitrous Oxide (N_2O).

d. Manmade carbon dioxide is estimated to be about 5 percent (1/20th) of the total CO_2 emitted. Animals and man are relatively minor contributors. Decaying organic matter is the major source, followed by volcanic activity and release from warmer oceans. Warmer water releases more CO_2 than cooler water due to decreased solubility of CO_2 with rising temperature. Many studies show that atmospheric CO_2 concentration rises AFTER warming, not before. So which is the cause and which is the effect?

e. Animals exhale CO_2 and breathe O_2, while plants use CO_2 and exhale O_2. Professional greenhouses often add extra CO_2 to increase growth rates. Increased plant growth removes much of the CO_2 released into the atmosphere. Between pre-industrial and present times, studies show an average of 15 percent increase in plant growth rates, with some species increased many times that, eg, young pine trees. Increased plant growth rates and wider distribution of arable (farmable) land due to warming as well as improved farming practices can solve the so-called overpopulation problem. If much of the data used in the climate models are based on proxy data from tree rings, and growth has been increased by CO_2, does that mean that the data is artificially skewed toward "warmer" results? Hmmm.

f. Critics created the "progressive nitrogen limitation hypothesis," which assumes that increased growth rates of trees would deplete poor soils of nitrogen, thus mediating the positive effects of increased CO_2. This is a scenario based on theory, not reality, which stubbornly refuses to support the hypothesis. Many studies[15] show that, contrary to the hypothesis, although roots grow deeper and produce more fine hairs, soil and forest floor are enriched in nitrogen from biological sources, ie, increased root mass and leaf litter supporting beneficial microbes in the soil.

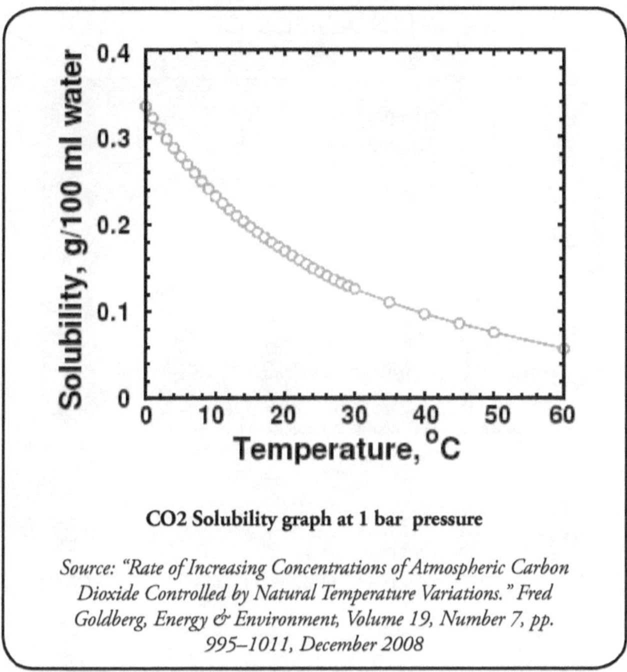

CO2 Solubility graph at 1 bar pressure

Source: "Rate of Increasing Concentrations of Atmospheric Carbon Dioxide Controlled by Natural Temperature Variations." Fred Goldberg, Energy & Environment, Volume 19, Number 7, pp. 995–1011, December 2008

g. One benefit of increased CO_2 is that the stomata (openings) of leaves, which take in CO_2 and emit water vapor and oxygen, are reduced, leading to less water loss, enhanced water use and improved tolerance to dryer conditions. At elevated CO_2 levels, stomata do not need to be open as far to allow sufficient CO_2 in for photosynthesis and, as a result, less water is lost through transpiration.[16] In controlled studies, an additional benefit of reduced stomata openings is a reduction of ozone damage.

h. The increased rate of growth of plants, from forests to sea algae, results in more of certain cooling aerosols being produced. These include Carbonyl Sulfide (COS) from soil and seas that become highly reflective sulfate in the stratosphere to reflect more solar radiation back into space; iodo-compounds[17] from sea algae that nucleate clouds to reflect more solar radiation back into space; and dimethyl sulfide (DMS), from seas that nucleates clouds and other aerosols such as isoprene from trees with similar effects.

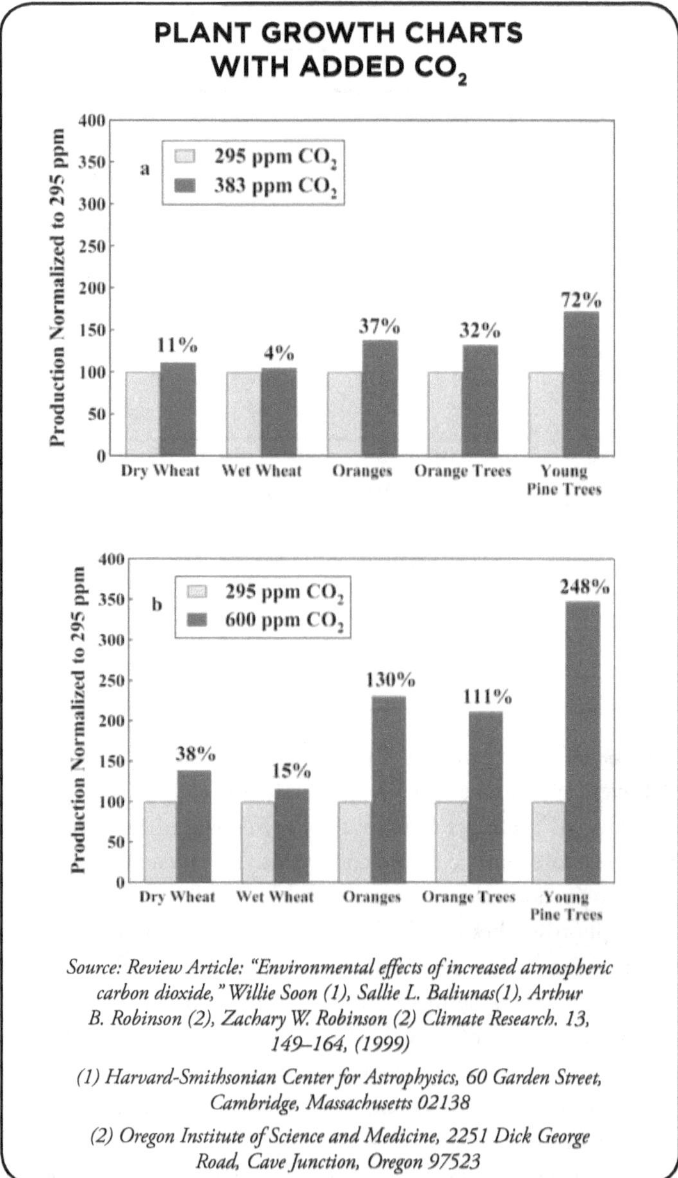

Source: Review Article: "Environmental effects of increased atmospheric carbon dioxide," Willie Soon (1), Sallie L. Baliunas(1), Arthur B. Robinson (2), Zachary W. Robinson (2) Climate Research. 13, 149–164, (1999)

(1) Harvard-Smithsonian Center for Astrophysics, 60 Garden Street, Cambridge, Massachusetts 02138

(2) Oregon Institute of Science and Medicine, 2251 Dick George Road, Cave Junction, Oregon 97523

The increase in carbon dioxide is greening many arid regions because of more efficient use of water and the increased growth rate. Sub-Saharan Africa is blooming, the Amazonian jungle is flourishing and global vegetative cover is increasing. The effect on ocean phytoplankton is equally as dramatic. Significant reduction of carbon dioxide levels as proposed by the various climate agreements would have a detrimental effect on plant growth and consequently food supplies.

Hormesis is a phenomenon, commonly seen in medicine and nutrition, where a low or moderate concentration or dose results in a positive effect, but a

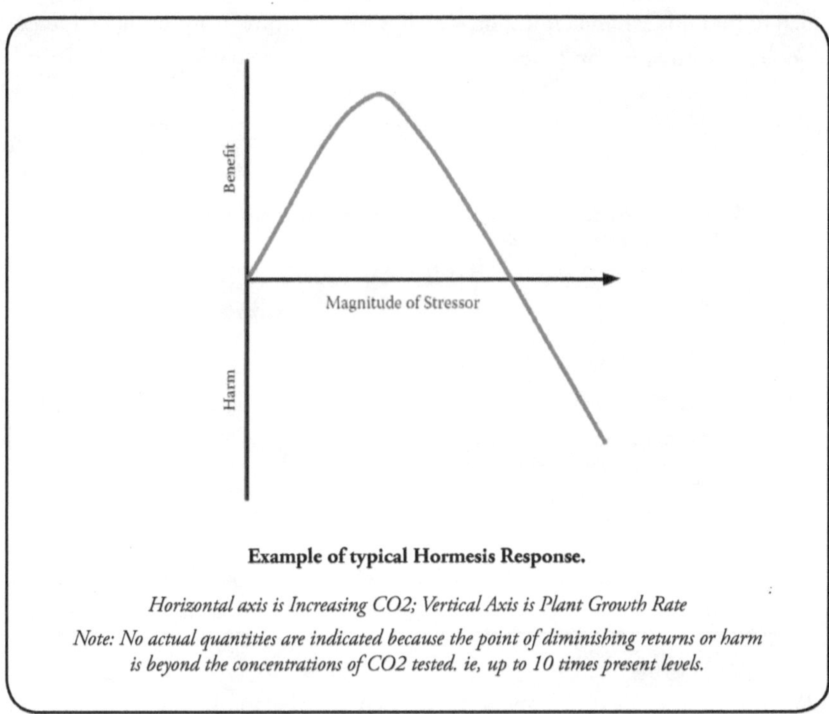

Example of typical Hormesis Response.

Horizontal axis is Increasing CO2; Vertical Axis is Plant Growth Rate

Note: No actual quantities are indicated because the point of diminishing returns or harm is beyond the concentrations of CO2 tested. ie, up to 10 times present levels.

larger dose results in damage. For instance, some salt and water are necessary to good health, but beyond a certain point, ingesting more can be harmful or fatal. The effect of CO_2 on plant life appears to be one such system. Increased CO_2 obviously benefits plant life, but it is uncertain at what level CO_2 might have a detrimental effect on growth. In professional greenhouses and experiments, even ten times the current level is still beneficial.

Claim 3. Carbon dioxide is important because it has a forcing effect on other factors such as water vapor

Truth: Since its absorption is already near saturation, very little additional forcing can take place due to increased CO_2. As stated above, water vapor doesn't stay as vapor to trap heat near the surface. It forms low altitude clouds that strongly reflect solar heat back out into space, overwhelming any re-radiation from the Earth. The models, which assume water vapor remains as vapor, predict an atmospheric "hot spot" at middle altitudes. Weather balloons and satellites have failed to find this assumed hot spot, which is the signature of atmospheric forcing of global warming in computer models.

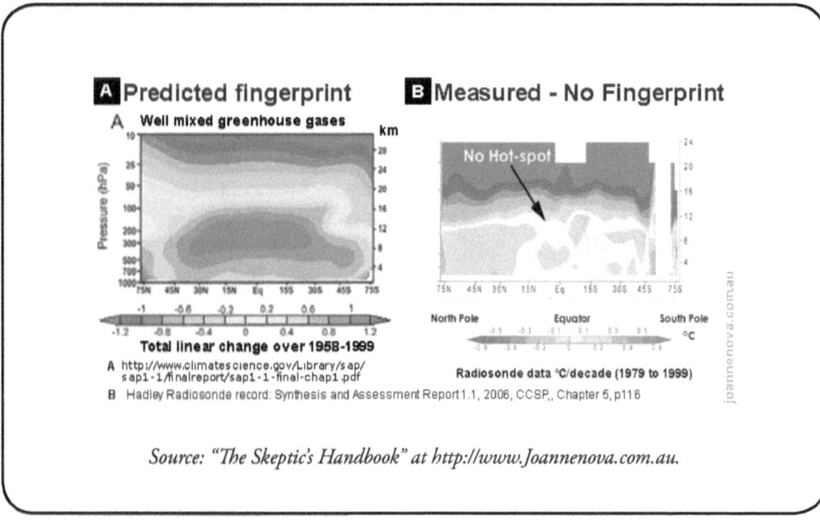

Source: *"The Skeptic's Handbook"* at *http://www.Joannenova.com.au.*

Claim 4. Manmade CO_2 levels have been rising rapidly due to increased industrialization and populations since the 1950s.

Truth: CO_2 levels have been steadily rising with warming. Recent increases in industrialization and population appear to have significantly contributed to the increase in atmospheric CO_2 since the 1950s when fossil fuel consumption began increasing. Rising temperatures have also contributed because of ocean release, so it is unclear how much is from manmade sources and how much is from natural processes. However, if CO_2 is not responsible for global warming, increased levels shouldn't alarm anyone and in fact should be celebrated as plant growth promoters. See 2c. above.

Claim 5. Developed countries are to blame for increased manmade carbon dioxide because they use most of the fossil fuels.

Truth: This is partially true. Industrialization is steadily increasing in developed and developing countries such as China and India. While developed countries are putting restrictions on themselves, international agreements exempt developing countries from such restrictions. These increases more than offset any gains from restrictions on developed countries.

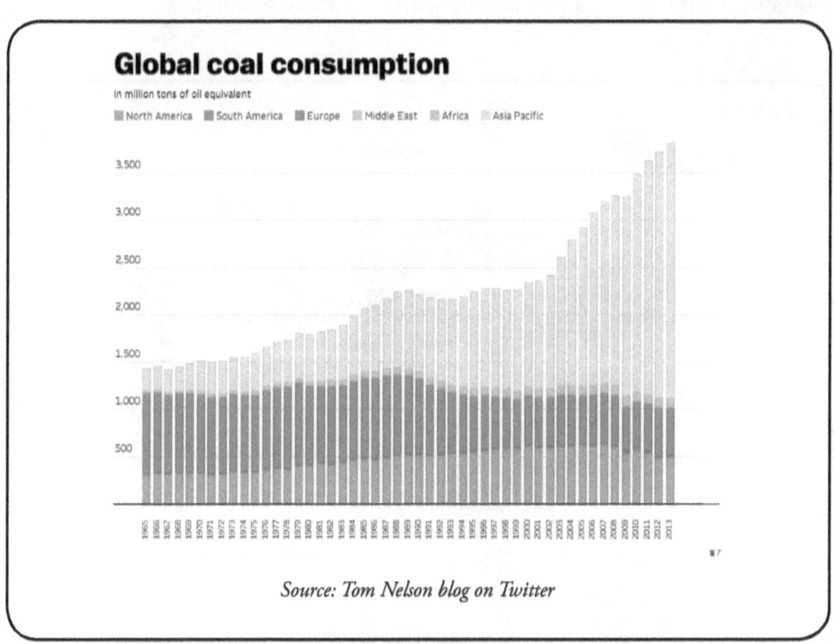

Source: Tom Nelson blog on Twitter

In addition to industrialization, increased cooking fires and subsistence agriculture to feed an increasing population are also significant contributing factors. CO_2 is CO_2. There is no escape clause for renewable sources. It doesn't matter whether it is from fossil fuels or burning dung or wood. Increased population in developing countries means more slash and burn agriculture and more cooking and heating by burning organic material. The modelers assume that renewable sources are exempt as causes because it is a renewable source. This is faulty thinking. Slash and burn agriculture of one acre releases a tenth of the carbon dioxide as ten acres.

Subsistence farming requires burning to release the nitrogen for crops. Modern agriculture releases far less carbon dioxide than subsistence farming, so keeping people in poverty makes no sense unless your aim is to stabilize or reduce the population in developing countries. It would be better if we helped developing countries develop modern agriculture and industry so they can clean up their act. When people are worried about how to feed their families, there is little time or incentive to do anything about pollution or the environment. In many underdeveloped countries the tradition of having as many children as possible is mostly due to the high rate of mortality in infancy and childhood from unchecked disease, poor diet and poverty. Without the incentive of high infant and childhood mortality, family size and populations could naturally stabilize.

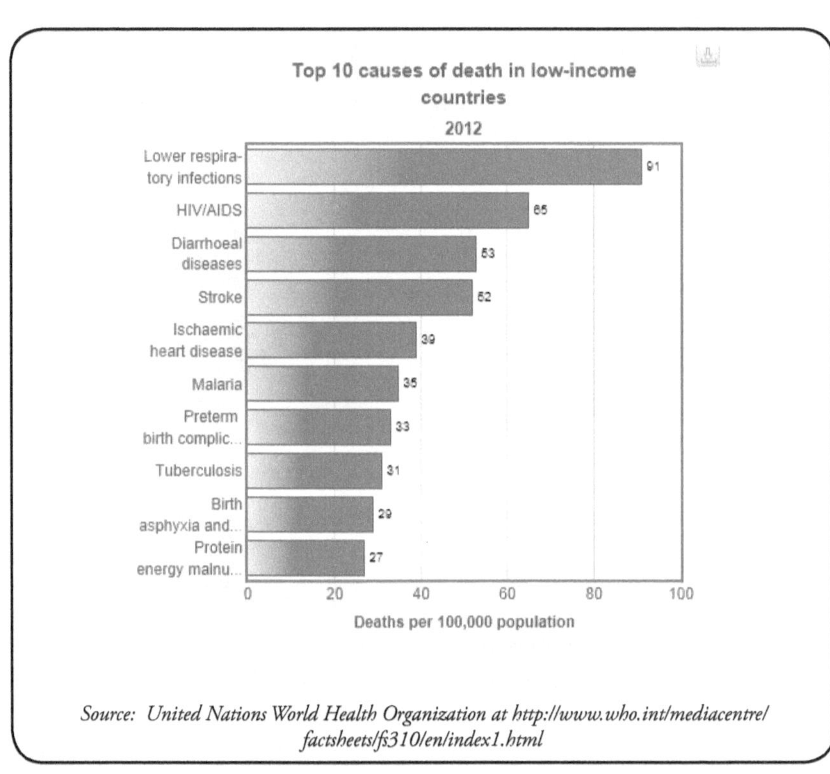

Source: United Nations World Health Organization at http://www.who.int/mediacentre/ factsheets/fs310/en/index1.html

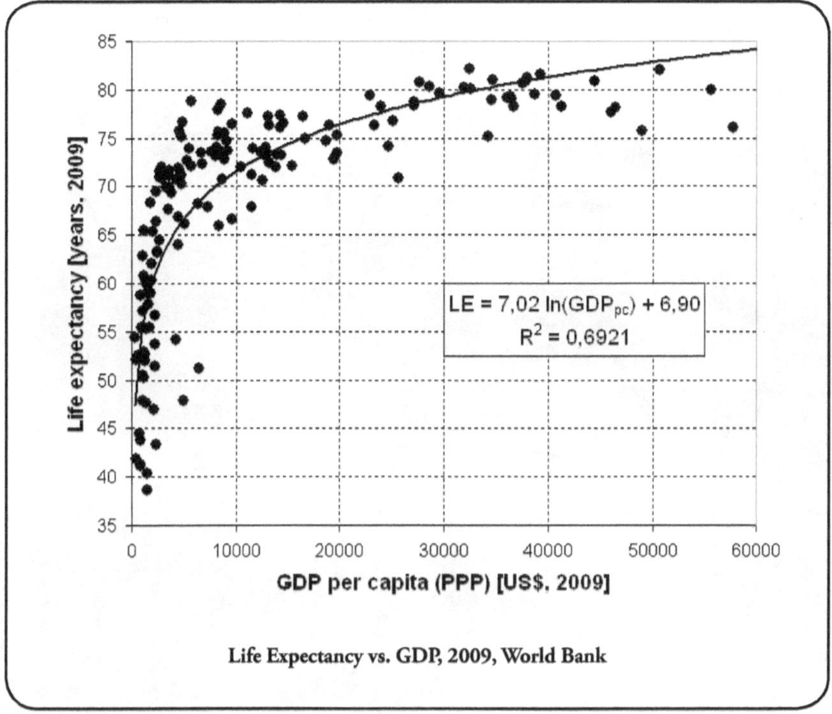

Life Expectancy vs. GDP, 2009, World Bank

Claim 6. Temperatures are hotter now than they have been in the last 100,000 years

Truth: This is clearly an unsubstantiated myth meant to scare people into compliance with drastic environmental regulations. The climate modelers have eliminated the Medieval Warm Period, which was hotter than it is today, and it was a time of prosperity. It was hotter in the 1930s than it is today. However, the American "Dust Bowl" of the 1930s was not due to warming. It was caused by opening up vast areas to farming that were poorly suited to it and a years-long severe drought. Based on historical accounts, ice cores and tree rings, modelers have dismissed the Medieval Warm Period and the Little Ice Age by claiming that they were not global phenomena but were limited to Europe and North America. More recent and more detailed ice core studies have shown that both of these periods were indeed global. Over the last 100,000 years, temperatures have been far hotter and far colder than the present. Who can say what "normal" global temperature is when it is always changing? Should we attempt to freeze the present day conditions as the ideal, or should we take a more reasonable approach to an ever changing climate?

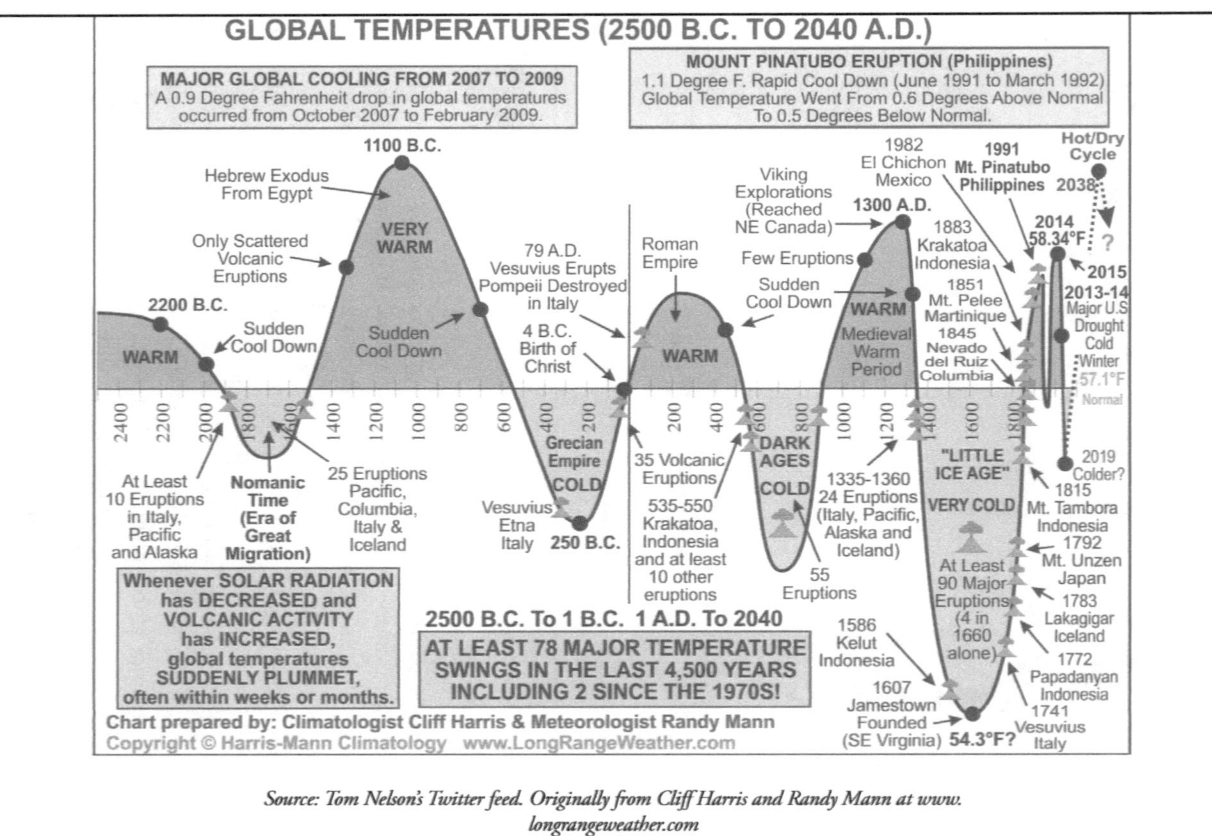

Source: Tom Nelson's Twitter feed. Originally from Cliff Harris and Randy Mann at www.longrangeweather.com

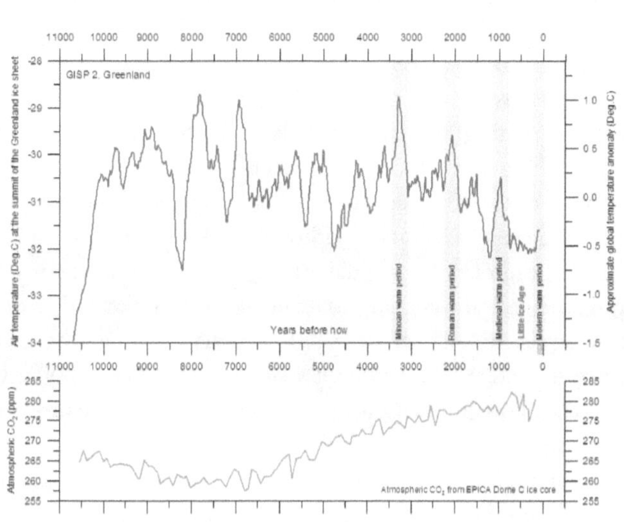

Temperatures since the last ice age compared to CO2 levels

This figure shows the climate swings since the last major ice age ended about 10700 years ago. The EPICA Dome C data ends in 1777[18]

Note that many of the temperatures are either higher or lower than those encountered in modern times, and that the CO2 levels do not correlate to warming or cooling during this period.

Source: Tom Nelson's Twitter feed

Claim 7. Temperatures have risen faster in recent years than ever before.

Truth: Remember that there is no such thing as a global temperature. It is the average of all of the reporting stations all over the world. For example, several years ago Ross McKitrick[19] showed that the rapid rise in temperatures in the 1990s directly coincided with a decrease in the number of Siberian weather stations reporting due to the break-up of the Soviet Union.

Additionally, the "hottest year on record," 1998, was an El Nino year so it was naturally hotter than the years just before and after. Another cause of rising average global temperatures is the urban heat island effect. Cities are hotter than rural areas. Many of the reporting stations that were once in undeveloped rural areas have experienced either suburban or urban development, or the stations have been moved to more urban settings. It is well documented that some have been, seemingly intentionally, relocated near or at heat sources such as paved parking lots and air conditioners.

One reason for relocation near buildings or other structures could be that new automatic-reporting equipment needs to be connected by cable. Rather than dig up parking lots or roads to install units in a grassy or protected area, many have opted to locate them where they can be directly connected without involving costly excavation, although such sites do not meet the stated requirements. Instead of excluding data from stations that are poorly situated, a convoluted mathematical algorithm (scheme) is used to "correct" it to presumed pre-industrial levels. In spite of all this, it appears that there has been no net warming since the late 1990s and even a slight cooling since 2005. One other problem with the new equipment is that it has a faster response time that records brief, transient signals such as car or plane exhausts that were not picked up by the older equipment. Because the equipment is designed to report maximum and minimum temperatures, this can create a false result.

Claim 8: The world is in danger of catastrophic consequences of global warming such as sea level rise, polar ice and glaciers melting, growing deserts, worse storms, droughts and floods.

Truth:

a. Sea levels have been rising along with warming since the Little Ice Age at an average rate of 7 inches per century due to glacier melting and expansion of seawater with warming. This rate has not changed significantly in recent times. To claim that sea levels have risen, the IPCC used a tide gauge in Hong Kong that showed tide levels rising, not because of sea level rising, but because the land is sinking (subsiding). Those islands that were supposedly in danger of being swamped have had almost no net sea level rise in all the years since the early predictions.

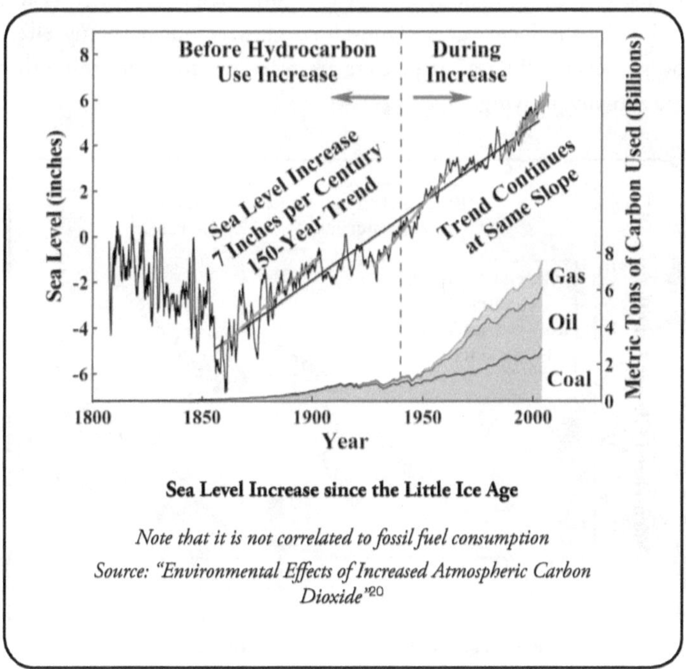

Sea Level Increase since the Little Ice Age

Note that it is not correlated to fossil fuel consumption

Source: "Environmental Effects of Increased Atmospheric Carbon Dioxide"[20]

b. Polar ice caps have shrunk and grown in recent years but overall they have remained relatively unchanged since preindustrial times. The media hype is about a theory that the Larsen Ice Shelf (in the more northerly Western Antarctica peninsula) might break away from Antarctica, causing rapid sea level rise of 5 meters (16.4

ft.). Although a part broke away in 1995, most experts say rapid collapse will not happen and any collapse and sea level rise would occur over centuries.[21] Both sea ice and ice cover have grown even more in other, more southerly locations on the continent. This year (2014) northern polar sea ice was thicker than usual so that there was some concern that the polar bears might have a harder time finding seals to eat. By the way, polar bear populations have been increasing in recent years.

c.   Glaciers have been receding since the Little Ice Age at a relatively steady pace that, along with water expansion with warming, accounts for much of the sea level rise. Melting of floating sea ice doesn't cause a rising sea level. It is already displacing its weight in saltwater. Ice expands as it freezes, so that melt water shrinks as it melts, resulting in the release of the same weight of water as was originally displaced. Only land-based glaciers will have any effect on sea level. While most glaciers are receding, there are some that are actually growing.

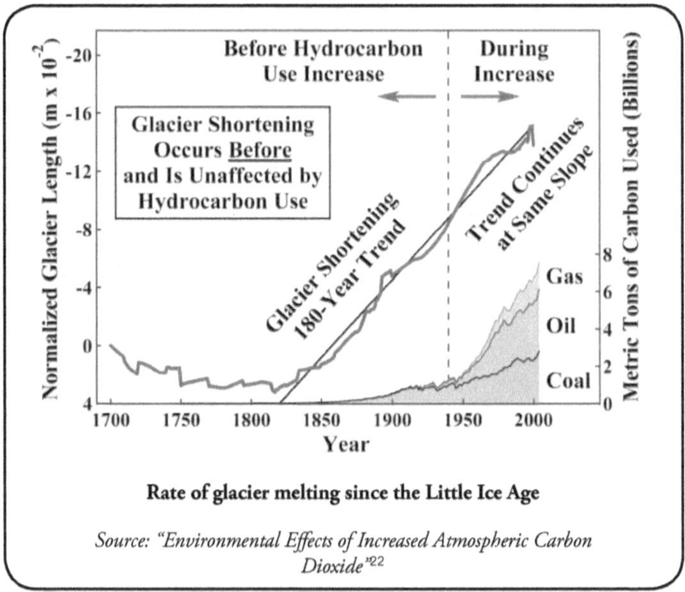

Rate of glacier melting since the Little Ice Age

Source: "Environmental Effects of Increased Atmospheric Carbon Dioxide"[22]

d.   Droughts have not increased in recent times. Some years are worse than others, but the overall picture has not changed.

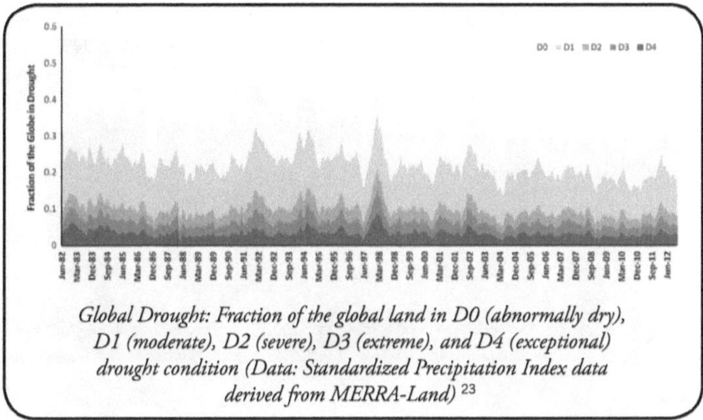

*Global Drought: Fraction of the global land in D0 (abnormally dry),
D1 (moderate), D2 (severe), D3 (extreme), and D4 (exceptional)
drought condition (Data: Standardized Precipitation Index data
derived from MERRA-Land)* [23]

e. Deserts are generally a result of geographic barriers and over grazing. Most of the Sahara Desert was once a grassy plain where livestock were grazed. It is a naturally dry area due to mountains to the west that block much of the moisture from the Atlantic Ocean. Overgrazing and loss of denuded top soil from winds had contributed to its expansion long before the industrial age. In most areas of the world, there has been no marked increase in the rate of desertification in recent times.

f. Storms have not gotten worse. The dollar damages in some areas have increased due to increased urbanization, but not the severity of the storms themselves. The number of tornados has actually declined in the United States and their severity has not increased. The same is true of hurricanes and in recent years few have made landfall in the United States.

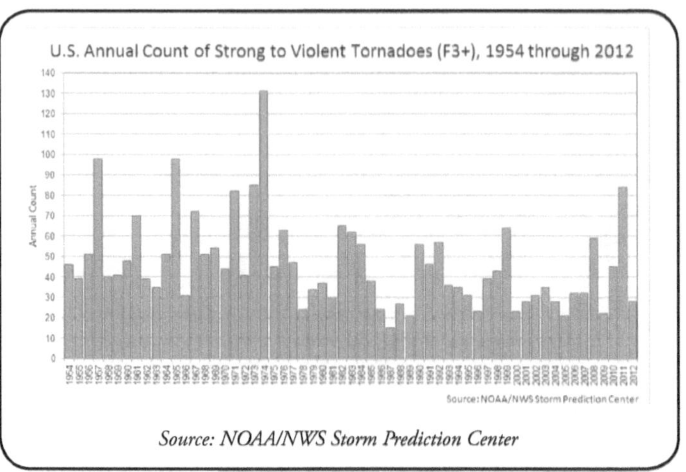

Source: NOAA/NWS Storm Prediction Center

Claim 9: The oceans are becoming more acidic due to the increased CO_2 and erosion of carbonate rocks by acidified rain, so that corals and other animals are being harmed or killed.

Truth: This is partially true, but not to the extent that is claimed. The oceans are naturally alkaline (non-acidic) and are buffered by minerals and by the activities of biological systems. (See below for explanations of pH and buffering.) The normal pH of the oceans is in the range of 8.1. Like the manmade global warming hypothesis and the progressive nitrogen limitation hypothesis, the ocean acidification hypothesis is a model scenario based on theory, not real experimental data. The models in the IPCC reports project a maximum reduction in pH of 0.3 by 2100. They also claim that a reduction of pH will slow the growth rate of coral or worse. So, are the corals and shelled creatures in danger? Hardly.

These models use a purely geochemical approach and do not take into account the effect of living systems on pH. For example, photosynthesis is known to increase pH. There is a temporary seasonal variation in pH from such processes that is greater than the projected pH changes from CO_2.[24] Real experiments by bubbled CO_2 through ocean water don't result in the same pH changes projected by the models. In these experiments, no net negative effect on coral growth is seen with actual acidification by adding hydrochloric acid. As a matter of fact, in these experiments a slight reduction in pH may have been beneficial.

CO_2 becomes carbonic acid (HCO_3^- and H_2CO_3) in water. $H_2O + CO_2 = H_2CO_3$, a very weak acid, some of which splits to form HCO_3^- ions and H^+. The free H^+ is what erodes carbonate ($CO_3^=$) rocks. For example, insoluble $CaCO_3 + H^+ = Ca^{++} + HCO_3^-$ (both soluble).

UNDERSTANDING PH:

pH is a measure of the hydrogen ion (H^+) concentration in a water solution. The more hydrogen ions there are the more acid the solution is. However, pH is the negative log of hydrogen ion concentration, so that the higher the concentration, the lower the pH number. For example, a pH of 5 is a hydrogen ion concentration of 10^{-5}. A pH of 8 is a hydrogen ion concentration of 10^{-8}. 10^{-5} = 0.00001 or 0.001 percent. 10^{-8} = 0.00000001 or 0.000001 percent. So pH 5 is 1,000 times the hydrogen ion concentration of pH 8. Neutral water is pH 7 with equal amounts of hydrogen ions and hydroxyl ions.

UNDERSTANDING BUFFERING:

When water is buffered, it has at least two related molecules or ions in balance that resist changes in pH of the solution by changing their ionic state when acids or bases are added or when diluted. Any added hydrogen ions from a more acidic solution causes a change in the ratio of the two buffer species offsetting any tendency to change pH. Water exposed to the atmospheric carbon dioxide is never neutral pH 7. It is in the range of pH 5.4. Oceans are in the range of pH 8 so it is obvious that the oceans are "buffered" against changes in pH by rain that is more acidic. Biological systems and a complex mixture of buffering molecules/ions contribute to the buffering of oceans and other bodies of water, so that pH remains relatively steady over long periods of time under various stresses from sources ranging from acid rain to volcanoes.

Example: CO_2 combines with water to form one of three forms of carbonic ions depending on the direction the pH is being pushed. These reactions can go back and forth between species as needed to maintain a constant pH.

$$CO_2 + H_2O \longleftrightarrow H_2CO_3 \longleftrightarrow H^+ + HCO_3^- \longleftrightarrow CO_3^= + 2H^+$$
$$\text{carbonic acid} \qquad \text{bicarbonate ion} \quad \text{carbonate ion}$$

The CO_2 to carbonate series is one of the most important buffer systems in the oceans and is supplemented by carbonate in its various forms from erosion of rock. This example is simplified for clarity by leaving out other ions that affect pH such as Calcium, Magnesium, sodium, chlorine and sulfite (Ca^{++}, Mg^{++}, Na^+, Cl^-, $SO_3^=$). Chemicals produced by living organisms also affect buffering and overall pH.

As stated earlier, CO_2 is less soluble in warmer water so temperature also affects pH. Warmer water means less CO_2 is dissolved in it by giving off more CO_2 to the atmosphere. A lower concentration of CO_2 with warming should, if anything, raise the pH (less acid) not lower it (more acid). However, due to buffering, no net change in overall pH has been found other than temporary localized variations from other processes like photosynthesis and agricultural run-off.

Claim 10: World governments must take drastic action now to prevent further warming and catastrophic consequences.

Truth: The jury is still out as to whether warming is a good thing or a bad thing. More people die from cold weather-related events than from heat. Warming periods in the past such as the Medieval Warm Period were times of increased prosperity and peace. There is no indication that any sort of a tipping point is approaching that would cause the predicted catastrophic consequences. It is not even certain that government action could have any effect on warming. If carbon dioxide is not the main cause of warming, then regulating it might be a fool's errand. Developed nations might curtail use of fossil fuels, but treaties proposed like Kyoto and Rio exempt developing countries. This includes China, India, and Mexico, among the largest and most industrialized developing countries in the world.

Claim 11. There is a consensus among climate scientists on the causes and dire consequences of global warming.

Truth: According to the climate change advocates, there is a consensus among climate scientists, the science is settled and the debate is over. But is that true? Not according to over 30,000 scientists who have signed a petition disputing their conclusions.[25] However, neither the "consensus" nor the petition has any meaning in science. Consensus applies only to opinion, not truth. Truth is never decided by a popular vote. It only takes one scientist with new (or ignored) facts to disprove the currently popular theory aka COWDUNG.[26] Remember that before Copernicus' new theory and Galileo's new facts the consensus was that the sun, and indeed the universe, revolved around the earth. The real scientific question should be, "What does the data say?" not how many experts believe or assume something is true. In the 1970s some of the same people who made the global warming models were predicting a coming ice age because of a short-term cooling trend and computer models that assumed a continuation of the trend.

"In short, under the new authoritarian science based on consensus, science doesn't matter much anymore. If one scientist's 1,000-year chart showing rising global temperatures is based on bad data, it doesn't matter because we still otherwise have a consensus. If a polar bear expert says polar bears appear to be thriving, thus disproving a popular climate theory, the expert and his numbers are dismissed as being outside the consensus. If studies show solar fluctuations rather than carbon emissions may be causing climate change, these are damned as relics of the old scientific method. If ice caps are not all melting, with some even getting larger, the evidence is ridiculed and condemned. We have a consensus, and this contradictory science is just noise from the skeptical fringe."

Terence Corcoran, in "Climate consensus and the end of science," Financial Post, June 12, 2006.

THE TRUTH ABOUT ANTHROPOGENIC GLOBAL WARMING

Climate change predictions are based almost totally on projections of computer models that assume that all factors of climate are known and their complex, often chaotic, interactions are really understood. It also assumes the current trend will continue. A projection is not a true prediction unless the computer model is 100 percent empirically correct and agrees with reality in every way. For a computer model to have any credence, every single factor must be thoroughly understood, the data on which it is based must be complete and beyond question, the correct mathematical formulas must be used, and unpredictable or chaotic behavior must be eliminated or managed. As stated in earlier chapters, computer models are only as good as the data used and the way it is analyzed. The rule is GIGO – garbage in, garbage out. There are many climate models used by the IPCC (United Nations International Panel on Climate Change), none of which agree on the degree of change although all project change based on the manmade CO_2 theory. The projections of the models differ from each other by up to 400 percent for the year 2100.

However, these models failed to predict the El Nino in 1997-1998, the dearth of hurricanes striking North America in recent years or the current stagnation of global temperatures. There has not been any global warming in the last 18+ years, which has the climate scientists baffled and worried that governments will fail to act in the ways that they have advocated. The models predict continued warming but the data show otherwise. Advocates refer to this as a hiatus, not a need to correct the models. Some have tried to say there is no pause in warming, but the data of even the staunchest advocates show there is. See the chart below for comparison of models vs. reality according to Hadley Climate Research Unit (HadCRU), University of East Anglia, UK, a leader in the warmist camp.

The models all assume that the future is predictable based on current data. Is that true? The truth is that we don't even know if we CAN predict climate in the future. The "predictions" are really projections based on previous data and a set way of analyzing it under the assumption that past trends will continue unchanged into the future in the same way that short-term weather predictions are done. When applied to the past, these models fail to postdict (predict into the past) what actually happened in the twentieth century when reliable data was available, much less the more distant past. The Medieval Warm Period certainly did not predict the Little Ice Age a few centuries later, nor did the Little Ice Age predict the continuing warming since that time.

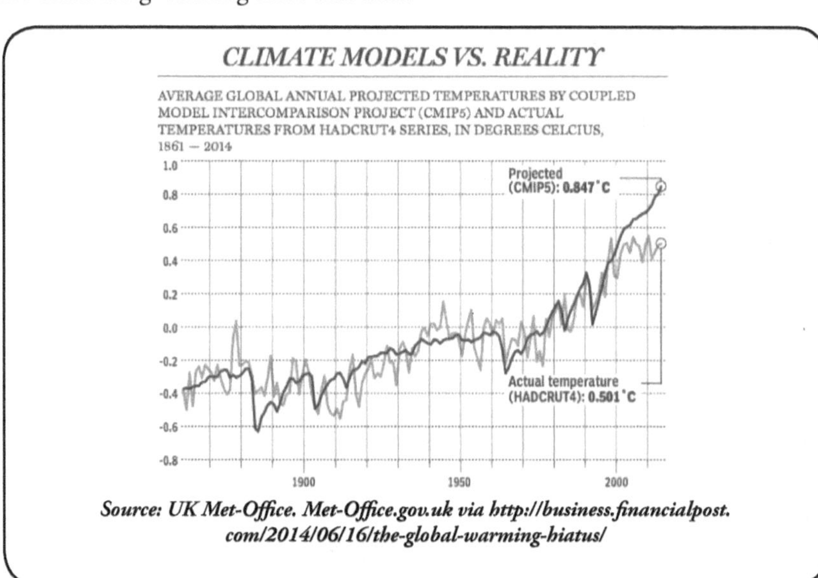

Source: UK Met-Office. Met-Office.gov.uk via http://business.financialpost. com/2014/06/16/the-global-warming-hiatus/

OTHER POSSIBLE CAUSES OF CLIMATE CHANGE

There are several other theories for climate change that fit better than AGW, listed as follows:

1. Solar variability
2. Cloud formation and albedo
3. Ocean current changes or cycles
4. Planetary motions

1. - 3. SOLAR VARIABILITY AFFECTS ALBEDO AND OCEAN CURRENTS SO THE THREE CANNOT BE COMPLETELY SEPARATED.

Global climate change, from cooling to warming to cooling, is strongly correlated with variable solar activity. The sun goes through periodic cycles of increased and decreased activity marked by increased and decreased sun spots. High activity means stronger solar wind; low activity means weaker solar wind. Cycles of 11, 87, and 210 years occur. Records of sun spots go back hundreds of years and correlate well to periods of warming and cooling. High activity periods (maxima) with more sun spots correspond to warming, but lower activity (minima) with few if any sun spots are associated with cooling.

The Maunder Minimum, from 1645 to 1715, was an extended period of low or no sunspots which marked the Little Ice Age. We have been in a period of higher than average solar activity as indicated by an above average number of sun spots. We have recently entered a period of relative quiescence with few if any sun spots. The expected return to a period of greater numbers of sun spots has been one of weakest Maxima ever recorded. That could mean we are entering a cooling period, and, indeed, there appears to be some minor cooling since about 2005.

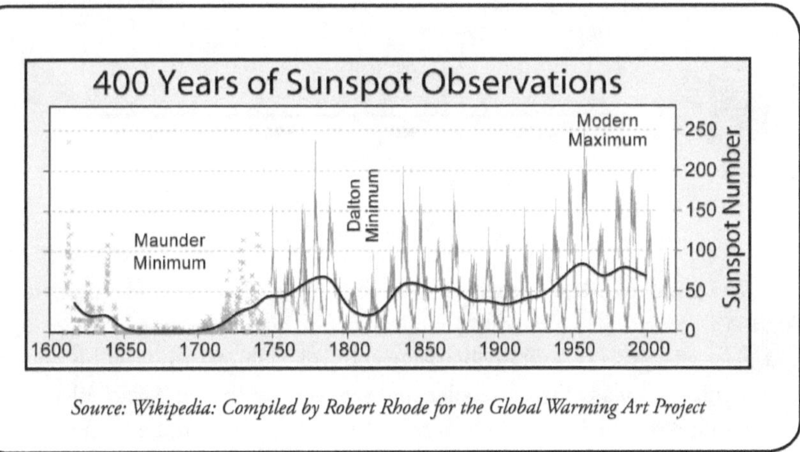

Source: Wikipedia: Compiled by Robert Rhode for the Global Warming Art Project

Cosmic Rays from the galaxy constantly bathe the solar system with high energy charged particles that are known to nucleate clouds and precipitation in our atmosphere. The solar wind, also composed of charged particles, and the associated magnetic fields, protect the earth from many of the galactic cosmic rays. Strong solar winds during high solar activity cause more shielding of the earth from galactic cosmic rays. Weak solar wind results in more galactic cosmic rays reaching earth and nucleating more clouds in the atmosphere. More clouds reflect more solar radiation back out into space, thus cooling the earth. Although it is not clear how, this cycle also affects ocean current oscillations such as the Pacific Decadal Oscillation (PDO) and its associated increased cloud cover and cooler temperatures. The Thermohaline circulation (THC), which mixes deep cold ocean water with warmer surface water, is another circulation cycle that may be affected by solar activity and cosmic rays and thus affect climate.

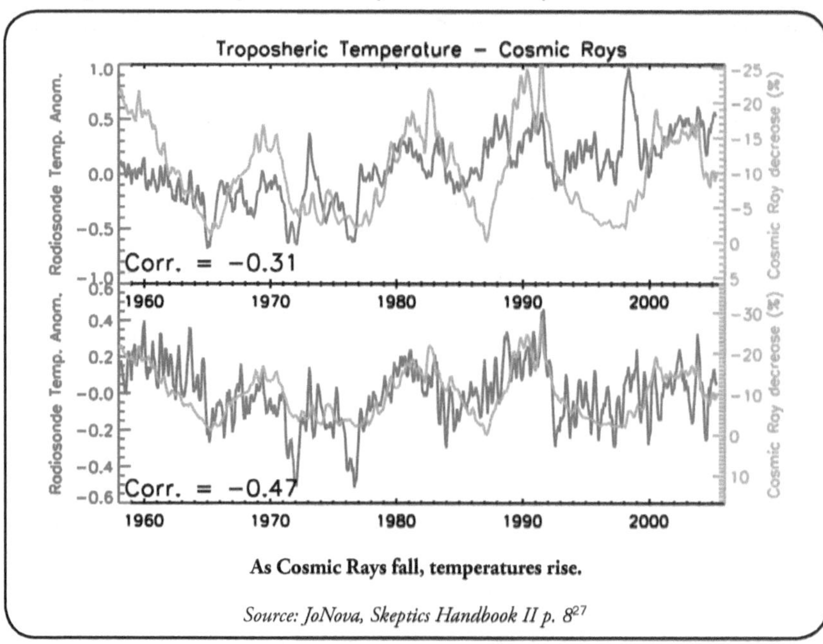

As Cosmic Rays fall, temperatures rise.

Source: JoNova, Skeptics Handbook II p. 8[27]

Seven independent records – solar activity (1); Northern Hemisphere (2), Arctic (3), global (4), and US(5.) annual surface air temperatures; sea level (6); and glacier length (7) – all qualitatively confirm each other by exhibiting three intermediate trends – warmer, cooler, and warmer. Sea level and glacier length are shown minus 20 years, correcting for their 20-year lag of atmospheric temperature. Solar activity, Northern Hemisphere temperature, and glacier lengths show a low in about 1800, (Dalton Minimum sunspot activity). Hydrocarbon use (8) is uncorrelated with temperature. Temperature rose for a century before significant hydrocarbon use. Temperature rose between 1910 and 1940, while hydrocarbon use was almost unchanged. Temperature then fell between 1940 and 1972, while hydrocarbon use rose by 330 percent. Also, the 150 to 200-year slopes of the sea level and glacier trends were unchanged by the very large increase in hydrocarbon use after 1950.

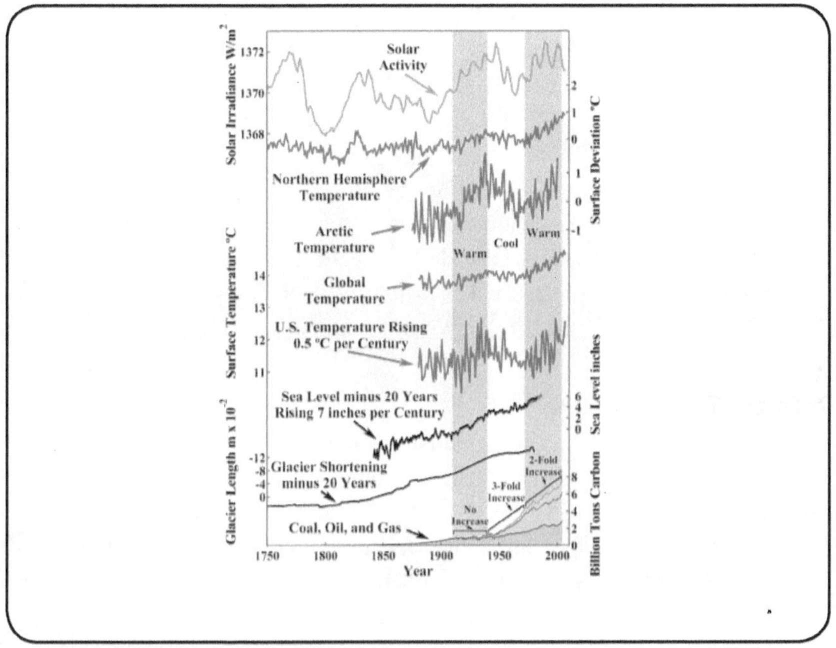

Source: "Environmental Effects of Increased Atmospheric Carbon Dioxide"[28] *Willie Soon (1), Sallie L. Baliunas(1), Arthur B. Robinson (2), Zachary W. Robinson (2) Climate Res. 13, 149-164, (1999); (1) Harvard-Smithsonian Center for Astrophysics, 60 Garden Street, Cambridge, Massachusetts 02138 (2) Oregon Institute of Science and Medicine, 2251 Dick George Road, Cave Junction, Oregon 97523*

REFERENCES FOR EACH OF THE 7 DATA SETS:

1. Hoyt, D. V. and Schatten, K. H. (1993) J. Geophysical Res. 98, 18895-18906.

2. Groveman, B. S. and Landsberg, H. E. (1979) Geophysical Research Letters 6, 767-769.

3. Polyakov, I. V., Bekryaev, R. V., Alekseev, G. V., Bhatt, U. S., Colony, R. L., Johnson, M. A., Maskshtas, A. P., and Walsh, D. (2003) *Journal of Climate* 16, 2067-2077.

4. National Climatic Data Center, Global Surface Temperature Anomalies (2007) http://www.ncdc.noaa.gov/oa/climate/research/anomalies/anomalies.html and NASA GISS http://data.giss.nasa.gov/gistemp/graphs/Fig.D.txt.

5. Jevrejeva, S., Grinsted, A., Moore, J. C., and Holgate, S. (2006) J. Geophysical Res. 111, 2005JC003229. http://www.pol.ac.uk/psmsl/author_archive/jevrejeva_etal_gsl/

6. Leuliette, E. W., Nerem, R. S., and Mitchum, G. T. (2004) Marine Geodesy 27, No. 1-2, 79-94. http://sealevel.colorado.edu/

7. Oerlemanns, J. (2005) *Science* 308, 675-677

8. Marland, G., Boden, T. A., and Andres, R. J. (2007) Global, Regional, and National CO_2 Emissions. In Trends: A Compendium of Data on Global Change. Carbon Dioxide Information Analysis Center, Oak Ridge National Laboratory, US Department of Energy, Oak Ridge, TN, USA, http://cdiac.ornl.gov/trends/emis/tre_glob.htm

2. PLANETARY MOTIONS

According to Milankovitch theory, variations in planetary motions affect Earth's climate over thousands of years. Whether or by how much any of this affects climate in the short term is not clear. Variations occur in orbital eccentricity, precession of orbital perigee and apogee (nearest and farthest distance) and axial tilt. Under the influence of Jupiter and Saturn, the eccentricity of the elliptical orbit of the earth around the sun varies from a near circle to an elongated ellipse every 413,000 years. Eccentricity varies from 0.000055 (nearly circular) to 0.0679 (highly elliptical). The present eccentricity is 0.017. Under the influence of the sun and moon, the Earth's tilt varies between 22 and 24.5 degrees from the orbital plane (ecliptic) in a 41000 year cycle. At present the tilt is 23.4 and is decreasing.

Both the axial tilt and orbital perigee and apogee (closest and farthest points) wobble (precess) in a combined 21,000 year cycle. At present, the nearest point of Earth's orbit to the sun (perihelion at perigee) occurs in January resulting in milder winters (nearer to the sun) and milder summers at aphelion (farther from the sun), in the Northern Hemisphere, but the opposite in the Southern Hemisphere. There, the tilt of the earth results in summers nearer the sun and winters farther from the sun. When the perihelion moves into July in about 5000 years, the seasons will be quite different from today.

CAN WE DO ANYTHING TO CHANGE THE CLIMATE?

Probably not. Even if the unlikely possibility of catastrophic climate change occurs, there is probably not much we can do about it. If CO_2 is not the culprit, and if manmade CO_2 emissions are only a tiny fraction of the total, then there is no climatic reason to curtail its emissions. And any savings on the part of developed nations will be more than offset by developing nations such as China. If the earth is indeed entering a cooling period, maybe curtailing CO_2 emissions could be counterproductive or totally ineffective. What else could we do? Not much.

If CO_2 is not important to global warming, then the only reason not to promote coal and other fossil fuel generation is other types of pollution that could be harmful to health or the environment. Developed countries already curtail pollutants such as fly ash, sulfuric acid, mercury and lead. Only developing countries are still belching out most of the pollutants from power plants. It would be worth our while to help them clean up their industries.

There is no shortage of oil and gas which are abundant and widely distributed. The assumption that we will run out of oil and gas in the near future is proving to be exaggerated or false. Peak Oil was a theory that said oil production would peak in the 1970s and then we were supposed to run out of oil and gas in the near future, which sent prices skyrocketing. With new technologies locating and safely extracting these natural products in ever increasing amounts in locations never before considered, these predictions have proven false. Some even theorize (speculate) that oil is a natural product of the Earth, not a fossil as believed for the last few centuries. If it is created within the earth by thermal and chemical action, then there should be no problem using what we need until technology creates a reliable replacement.

Wind and solar power generation are just in their infancy and have many major problems that must be overcome. This is not to say that we shouldn't continue to research these areas. Other safe and clean energy sources that are available today are hydroelectric, geothermal and nuclear.

THE TRUTH ABOUT RENEWABLE ENERGY SOURCES

Hydroelectric generation of power is the cleanest and best source at the present. It is low tech, consistent, low maintenance and clean. Its only down side is that it is hated by environmentalists and is being maligned as disruptive to the environment. They advocate tearing down existing dams, not building more for more power. Most of the existing dams offer flood protection for people downstream. Without the dams, catastrophic floods would return.

Nuclear reactors are the second best source of clean, safe energy. It is high tech but has no polluting emissions. Even the cooling waters are not released into the environment until they are cooled to match the temperature of streams. Photos of cooling towers with steam coming out are used by environmentalists to imply pollution. Since only quickly cooling steam is emitted, the most it could do is promote clouds and plant growth. Modern designs are much safer than previous ones, but due to government overreach and clamorous environmental protestors, there has not been a new plant built in the US in nearly 20 years.[29] After being halted in 1988, construction of Watts Bar 2 reactor was resumed in April 2012 and will be completed by 2016. Two other existing nuclear plants, Summer and Vogtle, are adding two new reactors each. These units are scheduled to be completed by 2018. Several older plants have been closed in recent years as being too costly to maintain. The only issue is with waste management that requires long term storage until it is feasible to recycle most of it. Waste can be safely stored in underground vaults in arid and seismically stable areas if politicians and bureaucrats would get out of the way. Yucca Mountain Repository was shut down by the Obama administration before it became operational with no replacement available and none planned. Meanwhile, waste is stored on site at each plant.

Geothermal Energy is good but its use is limited to the few geothermally active areas available that are not prone to earthquakes. It uses heat from the earth to create steam that turns a conventional steam turbine so it is low tech and low maintenance.

Wind Energy, as it is practiced by power companies today, uses vast areas of multiple turbines on high towers with huge blades that need high winds to move them. When the winds are too low, power is not generated. When winds are too high, turbines can catch fire or blades can snap off or distort, making this form of energy generation intermittent, unreliable and in need of back-up power plants. The problem with this is that cranking up a hydrocarbon fueled power plant requires more than flipping a switch. It can't be done quickly or economically and it is hard to predict when and how much backup will be needed. Even if all of these problems can be overcome, these huge windmills create a vacuum behind them and kill large numbers of large and small birds by sucking them into the blades. Anyone who has lived near these wind farms will tell you that the low frequency sound and infrasound (too low to hear but that can be felt) are annoying at best and debilitating at worst.

Solar Energy 1: Photovoltaic cells are high cost, high tech, require exotic chemical elements and sophisticated manufacturing facilities. They are also low efficiency (less than 30 percent), require vast areas to generate enough power to be economically feasible, have a short lifespan, depend on low cloud cover making them inconsistent, so that they require backup hydrocarbon fueled power generation.

Solar Energy 2: Solar mirrors aimed at steam generators are high cost to install, but use low tech mirrors and conventional steam generators. They require vast areas of mirrors and depend on low cloud cover making them inconsistent, so that they require backup hydrocarbon fueled plants for generation. If that is not bad enough, the mirrors kill birds because of the high heat they produce. They literally fry birds in flight and the light blinds pilots flying nearby.

Ethanol fuel is more polluting than gasoline to produce, uses food crops that could feed starving millions, damages engines and is a lower mileage fuel than gasoline. Given government subsidies for growing corn for fuel, many farmers abandon more needed crops such as wheat and soybeans to plant more corn. Meanwhile, the price of food increases and many in the world go hungry.

Electric cars produce more CO_2 than conventional internal combustion cars because they use batteries that must be recharged from power stations hooked up to power plants, many of which are coal fired. It is more efficient to directly burn gasoline than to convert coal or even natural gas to electricity and transport it many miles over high voltage lines. It is a fact of physics that there is always a loss in efficiency at each stage of production and transport. For electric cars to reduce fuel consumption they must be recharged from solar, wind, nuclear, geothermal or hydroelectric sources.

WHY CLAIM CO_2 IS THE CAUSE

We have seen from the data that CO_2 cannot be the major cause of global warming (aka climate change), and that water is the most important greenhouse gas for regulating temperatures. We have also seen glimpses of other causes that have not been considered. So, two questions arise. Why all the attention by international bodies to CO_2 while neglecting or dismissing water's influence and other possible causes of global warming and cooling?

At its founding the IPCC of the UN was tasked to find a link between human activity and climate change, and so, like most political bodies, they did not look for natural causes. CO_2 is the one component that would favor government power and control of man's activities, while crippling industry and creating a guilt-filled picture of development. The aim of the IPCC and others advocating AGW is not saving the planet from climate change, but really is imposing socialism and anti-capitalism.

Please don't get me wrong; many people working in the field are sincere and truly want to make a difference for the planet. They believe the "party line" and work in a way that supports the erroneous premise without questioning its validity. That is the value of the so-called consensus – to draw in those who don't realize it is a scam or hoax. Sorry to insult you if you are one of those. It has nothing to do with your intelligence or credentials. Very intelligent and well-meaning people can become what are called "useful idiots" in the communist playbook.

The UN is overwhelmingly socialist, communist and totalitarian. As such, they hate the free enterprise and free markets of democracies that have led to the greatest technological and economic advances in history. They are not comfortable without controlling everything. Freedom is frightening to them. That is why they have fought so hard for draconian regulations to be put in place by developed nations immediately without hesitation or questioning the validity of the premise or counting the economic devastation that would ensue.

The aim of the UN has nothing to do with what is good for the planet or mankind, particularly in poorer countries. Some of the critics of AGW cite redistribution of wealth to poorer countries as one of the aims. While this is a part of the picture, it is a minor one compared to the desire to destroy technology, industry, free enterprise, free markets and freedom in general. Redistribution to poorer countries, if you listen to their statements, is intended to help them adapt to climate change, not to build their infrastructure or industry or economy. It is to keep them poor and primitive, with limited life spans, disease and deprivations. The money spent by the UN and others would be better spent in bringing developing countries up from their poverty, not by a hand-out but by economic and infrastructure development.

The aim of many of those working in the climate change field is also about money and power. Organizations that work on the AGW project, whether as researchers, bureaucrats or energy companies, all rake in billions of dollars if they support the premise. Those who do not support the "consensus" get excluded from grants and publication by activists in the UN, the US and other governments, academic and scientific organizations and publishers. People whose jobs or next paycheck depend on their support for the scheme are likely to find a way to support it, even if it means ignoring other data or phenomena that call the premise of AGW into question. It is unlikely that most people involved would jeopardize their cushy positions by bucking the trend.

Congress, for instance, gets what it pays for, and that is a major problem. Contrary to what the press would have you believe, government funding is no more unbiased than industry funding. If your present and future funding depends on finding a duck that Congress wants, you'll find a duck, even if all the data says it's a chicken. In general, you find only what you look for—and what your sponsors expect. This is the problem with government funding of any type of research. It skews the results toward the favorite conclusion of congress, whether it is correct or not. And let's face it, most politicians might be good at politics but are dumb as a box of rocks when it comes to real science. They are easily swayed by activists, popular causes and media hype. The search for empirical truth suffers from bias in favor of getting that next government grant and getting published in prestigious, though biased, scientific journals.

Aren't we safe from the demands of the AGW crowd and the media if we are in a cooling mode that does not agree with the AGW models? No, they are still shouting about warming and sea level rise and worse storms and drought as if nothing has happened. All they have done is acknowledge that there is a hiatus or temporary pause in warming. The drum beat is still the same. They proclaim that it will get hotter, a tipping point is coming and we are all in danger of catastrophic consequences. Because the agenda is really about controlling everything and growing governments, data and real conditions don't matter. Even if we enter another Little Ice Age, they would find a way to still blame it on human activity and demand immediate government action. Those with this agenda never give up and never acknowledge mistakes. They just move on to the next way to attempt to manipulate people. To them, the problem is never the real problem and the supposed agenda is never the real agenda.

> "Geese are but Geese tho' we may think 'em Swans; and Truth will be Truth tho' it sometimes prove mortifying and distasteful."
>
> *- Benjamin Franklin*

The bottom line is that any real science is used, distorted, massaged, faked and exaggerated to further the political and ideological agenda to control people, grow governments, and to cripple and disable development and real progress.

PART 8

SUMMARY AND CONCLUSIONS

CHAPTER 16

SUMMARY

I have only presented the heart of four foundational theories, evolution, origin of life, cosmology and quantum physics, along with the more recent climate change debate as a modern example of an agenda driven philosophy disguised as science. Many more topics can be presented that are based on similar beliefs and assumptions, but these four are sufficient to illustrate the problem.

Remember, **science is the pursuit of truth about the predictable, repeatable and measurable aspects of the universe with which we can or could conceivably interact.** One-time events such as evolution of a new species, origin of life and the origin of the universe are not repeatable so they cannot be fully examined by science. The best that can be done is either forensics, whereby attempts are made to piece together past events from evidence, or, lacking sufficient evidence, deductive reasoning to theorize about what must have (read is believed to have) occurred in the past to produce our present reality. It is OK to theorize about unknown and unknowable events, but it is not OK to claim such theories are actual facts, regardless of mathematical calculations or computer models.

Evolution taught others how to present pseudoscientific philosophies as science by using political tactics to unseat established leaders and to exclude God, either as a first cause or a cause that set up the laws making evolution possible. Given enough time, great complexity would naturally arise, a step at a time. Natural selection would eliminate inferior species while preserving superior ones.

Origin of life theories made it appear that life could arise from non-living chemicals. The theory that molecular evolution would produce living beings from nonliving chemicals made it seem that life was possible without design or a designer. Neither concept has ever been shown to be true. It is a myth with no basis in truth and it is magical thinking. The probability of forming even one specific functional protein or nucleic acid from previously formed, readily available amino acids or nucleotides[1] is so small that the event would be considered probabilistically or statistically impossible. Even if that occurred once by chance, it is not logical to believe that such an improbable event could happen, all in the same place, the hundreds or thousands of times necessary to form each of the specific proteins in sufficient amounts needed for the simplest of life forms, much less assemble them in the right order and quantity. In a simple example, if you drew an ace of hearts from a deck each time for countless times, you would logically be correct in assuming that someone was rigging the game. It is a gross understatement to say that the undirected origin of life by chance is many orders of magnitude more improbable than that.

Cosmology presented a picture of a progressive universe that invented itself and evolved without any need for direction. The Big Bang is based almost entirely on three things: 1.) a preferred interpretation of redshift of the light from distant objects as recessional speed, 2.) one preferred solution (of many) of Einstein's field equations that predicted an expanding universe and 3.) interpretation of the cosmic microwave background (CMB) as far redshifted light from shortly after the Big Bang beginning. The Big Bang made it impossible to determine a first cause because nothing could be determined from before that event. According to the theory, at the Big Bang time didn't exist, so the concept of time before that event would be meaningless. Even "before" would be meaningless. It short circuited the infinite series argument for a necessary first cause without addressing what caused the Big Bang or why there is something instead of nothing.

Quantum physics, with its interpretation of subatomic events as indeterminate, removed the need for the first cause. The universe could be said to only *look* deterministic based on mass effects of indeterminate foundations in the same way that life is said to only *look* designed based on countless evolutionary changes over vast periods of time. Cause and effect would be meaningless if particles are believed to exist in a multiplicity of states at once, if they only capriciously choose a random state when measured and if time and processes are completely reversible. If the universe is indeterminate at the subatomic level, then there is no cause and effect and no first cause, aka God.

Climate change, though similar in design and delivery, is different from the other areas discussed in that it has an immediate socialist agenda to destroy civilization's real progress based on bad science while ignoring real data. This is also true of the entire environmental movement, which is used to destroy real progress and impose a socialistic system. Green is the new Red. Behind both environmentalism and the climate change debate is a socialist agenda of control. Like the other areas discussed, it is a progressive philosophy in which the agenda is more important than the truth or any evidence to the contrary. It is based on a type of hubris that claims we not only completely understand our world but can re-engineer the climate and, indeed, the world. The truth is that we understand very little of such a complex and often chaotic system. We probably couldn't change it if we tried, beyond controlling individual local pollution sources with regulations. Would we really want to change interwoven processes of the entire planet without first being certain that there are no unforeseen consequences? Various aspects of environmentalism will be discussed in a future volume.

CHAPTER 17

WHY PROGRESSIVES BELIEVE AS THEY DO

What could motivate both the originators and their audiences to perpetrate such anti-science beliefs? Let me preface the answers by saying that many or most of the practitioners in the sciences and science education are not aware of the nonscientific underpinnings of their craft. These underpinnings do not invalidate their real research, only its interpretation to fit the accepted nested set of assumptions. They are merely working within the paradigm in which they were educated. Because all of their textbooks told the same story, they have assumed that the paradigms were proven beyond a doubt in the past, and have accepted them at face value. There is also an undercurrent of groupthink whereby scientists validate each other based on their mutual publications and opinions. They have not actually investigated for themselves because they assume the subject is beyond question. For my first degree in biology, I was once one of these, although I never conceded that God had not directed evolutionary developments.

A few, especially in academia, are actively pursuing an anti-God, atheist or Progressive agenda. Of the many areas where science has been replaced or corrupted, the same or similar agenda driven ideologies are the basis. It is a belief in Progressivism and a desire to do away with the necessity of God as the ultimate or first cause and consequently the Christian religion. Progressivism is a very old idea that originally was only about man's knowledge and society continually progressing, built on accomplishments of those who came before. The philosophy of Progressivism itself has been broadened from a reasonable belief in the progress of knowledge and social systems and has been reinterpreted into a magical belief that the universe, life and society are naturally progressive, going from simpler to more complex and progressing toward perfection without the need for direction by any higher power.

As a religion, Progressivism has been adopted by the materialist, socialist and communist philosophies to mean that society and man's very nature are moldable. Their mission, which gives their lives meaning, becomes helping that progress happen, based on their own understanding of perceived needs and their ideas of a perfect future. In the communist philosophy thesis followed by antithesis leads through struggle to synthesis in the form of a dictatorship of the proletariat and ultimately, to utopia in which both social ills and government fade away. However, communism never progresses past the dictatorship stage and ultimately falls apart even with strict controls. Progressive is a gross misnomer as applied in these social movements since it has always resulted in Regression into loss of freedoms and increased poverty. Progressive is really regressive.

As applied to science to justify anti-God theories, it results in fantastical schemes and magical thinking. As a matter of fact, the whole concept of a naturally undirected progressive universe is magical thinking with no basis in logic or reality. According to the second law of thermodynamics, increasing entropy is the rule. Over time, chemical and physical systems become less organized and naturally move to their lowest energy and organizational levels. Reduced chaos and increased complexity or energy both require work through input from outside the system.

Why are progressives, who believe in Darwinian evolution, trying to stop evolution (aka development) in its tracks? Why are progressives, who supposedly care deeply about impoverished peoples, so eager to keep them that way and drag the rest of us down to their level? Progressivism is a religion that allows followers to believe they have superior minds and intentions, that they know what needs to be fixed and that they can and should improve their world. It gives their lives meaning, which is lacking without acknowledging the necessity of God.

In truth, those who think they are smarter and morally superior are really engaging in magical thinking that is common in adolescence, immaturity or shallow thinking. It arises from underlying feelings of either inferiority or egotism, which itself is often based on deep seated but hidden insecurity. Those who have to think of themselves as better than the average person in order to feel good about themselves, are truly insecure, needing their ego stroked by adoring masses to validate their self-worth. They are also easily bruised and defensive about their self-worth and stature. That is why it is difficult to talk to progressives about anything that is not in complete agreement with their mindset. They are sure they know best, automatically put you in an inferior class and are offended by your impudence in questioning their beliefs. They may even attribute your failure to agree with them to some sinister motives and/or lack of intelligence.

They often cannot compromise or even listen to any ideas that challenge their sense of control and exclusive moral superiority.

They love to pontificate and personally attack as inferior or sinister anyone who doesn't validate their exalted opinion of themselves. Their favorite game is "Gotcha!" in which they make unfounded assertions in an attempt to make their critics look or feel foolish and put them on the defensive. In this climate, most people avoid such confrontations because they find conflict very unpleasant. It is an effective way of silencing critics.

CHAPTER 18

WHERE DO WE GO FROM HERE?

THE PROBLEM:

A progressive, anti-God mindset has come to dominate science and many other aspects of our world. Magical thinking, hidden agendas and protection of turf have replaced the need and the pursuit of truth.

The progressive monopoly relies on three things: academia, media and government funding. Academia, rather than being the open forum for discussing diverse ideas it should be, has become a closed priesthood where dissenting opinions are repressed, punished and blocked from funding and from academic publication by biased editors and peer reviewers. The popular press, whether print or electronic, is gullible and all too eager to parrot the standard line of SQR (Status Quo Regurgitators) without question. This includes the entertainment industry. Government funding grants aimed at supporting the standard line and controlled by SQR department heads, control what research can and cannot be done.

SOLUTIONS:

A public awareness of the problems is needed, leading to an open debate and holding accountable those blocking the free pursuit of truth. More, not less, research and discovery would result. Daylight is a good disinfectant.

- It will be very difficult to break through the progressive monopoly and inform the people of the problem, so it may take a generation. That is why books like this are important. It is one more way to chip away at the imposing edifice of the Progressive monopoly.

- The public must demand that academic departments allow and encourage questions, debates and alternative research.

- Peer reviewed publications must be held accountable for bias and advocacy that blocks publication of valid research based on alternative views.

- The popular press and the entertainment industries must be held accountable by an informed public for biased reporting and propaganda disguised as entertainment.

- Government funding should be minimized and scrutinized to assure that meaningful research is done and that outcomes reflect the unbiased truth.

Educate yourself so you can use logic and facts, not emotional appeals, in the debate. Arguing based on the Bible won't help your case with those that don't believe; it will only give your opponents ammunition to claim that you are anti-science and superstitious. Stick to facts and logic. Find and support those publications, websites and blogs that point out bias and give voice to alternative views as one way of educating yourself. Be discerning about internet information sources so you don't get misled by illogical, unsupported wild speculation on either side. Learn to recognize the difference between opinion and facts. Start blogs of your own to give others an outlet for dissent.

TACTICS – DIRECT CONFRONTATION:

This new challenge for each of us to reveal the truth won't be easy. As a matter of fact, because of the typically harsh and defensive responses of the progressives, it will sometimes be downright unpleasant. Most people will avoid confrontation on those grounds, preferring to avoid conflict and remain silent. Many are aware of the problem, or at least some part of it, but few are bold enough to stand up to the elitist bullies who call them names, disparage their character and engage in hate-filled diatribes. Remember, it's not about you. It's about them defending their fragile egos, their territories and their control of the situation. They will try to make you defensive and emotional like they are by attacking your character and intelligence. The key to discussing anything with them, assuming they are at all willing, is to remain calm and logical, not emotional (offended, hurt, angry). Don't let them put you on the defensive. Stand your ground and continue to ask your questions and make your logical points in a relatively nonthreatening manner. As long as they don't actually hit you, their tantrums and accusations mean nothing. If they manage to get you to address the issue emotionally, you have lost at least half of your argument. Not only will you be mischaracterized, but in any argument (discussion) when emotion enters, logic goes out the window.

For those of you who are confident enough, I would encourage you to enter those very debates without fear or dread with business-like or diplomatic composure. It may be useful to start with quiet discussions on milder aspects of their beliefs in private so they do not feel obligated to defend themselves in public. By all means, don't try to defend yourself or give them counter examples to illustrate that you are not guilty of whatever their accusations are. You would be wasting your breath, falling into their trap and giving them ammunition to shoot back at you. By putting you on the defensive they have achieved half of their goal of sidetracking the discussion and making you look foolish.

Let the vitriol flow over you without obvious effect. Address the facts and the possibility that their thinking might have some logical flaws or blind spots. When they lose their composure or spew slanderous remarks, that is a sign that you are on the right track because they are attempting to use intimidation to shut you up. Remain calm and repeat your point in a gentle, composed manner. Point out the logic of both points of view without sinking to their emotional level. Whatever the outcome, thank them for being willing to discuss the issue.

If you are a student, you may have to repeat the status quo line and not be too outspoken until after you get your passing grade. Remember, the professor is protecting his image to the rest of the class. In those cases, it is probably best to wait until the class is over to raise your questions in private and in a relatively non-threatening way if at all. You will have to gradually judge their tolerance for dissenting voices to see how far you can go in educating them. Some have such fragile egos that you will not be able to engage them at all. They may be unwilling to discuss their beliefs, so some finesse may be required to keep the door to communication open. Some, however, will be at least somewhat open to discussion of their beliefs.

Department heads or deans need to hear from students about those professors who are abusing their positions to proselytize or indoctrinate instead of teach. If no one complains, these leaders really don't know what is going on. One complaint may not have any effect, but complaints from numerous students may have some effect in reining in such abuse. Talk among other students and compare notes about the professor's behavior. Encourage others to also complain, preferably after the course is completed.

For parents, grandparents, and older adults, there is another way. You can enroll and audit classes. Most institutions of higher learning waive all or most of the fees for people over a certain age, eg, 65. To audit a class, you would sit in on classes but would not be graded or get college credit. This would be especially useful for subjects with which you have some prior knowledge. As an auditing student, without fear of failing the course, you can ask thought-provoking questions and discuss points not well covered by the professor. However, it would not be a good idea to audit a class in which your own child or grandchild is enrolled. It might jeopardize his or her grade.

TACTICS – INDIRECT CAMPAIGNS:

Start contacting those who control the message and the money. Write letters, emails, comment on social media, websites or blogs and phone those who can affect change. Contact members of Congress, heads of government departments, academic deans and heads of academic departments, media producers and editors, as well as advertisers in biased media. Let them know that you do not accept the bias that is exhibited. Give logical arguments for free exchange of ideas and discussions of logical alternatives. One person contacting them may not have much effect, but numerous contacts cannot be ignored. Most of these leaders and politicians assume that one letter from a proactive person represents opinions of many others not willing or able to speak out.

APPENDIX A

ANALYSIS OF
ON THE ORIGIN OF SPECIES

Analysis of *On the Origin of Species by Means of Natural Selection or The Preservation of Favoured Races in the Struggle for Life*, first edition, Charles Darwin, John Murray, London, 1859; Dover Thrift Editions, Mineola, NY, 2006, an unabridged republication of the original work.

Since it would be too lengthy and complicated to present the analysis of the entire text, I have chosen to present analysis of the original Introduction and Chapter XIV, Recapitulation and Conclusion. The entire text of each paragraph of the Introduction will be included followed by an analysis, and then a brief critique of Chapter XIV Recapitulation and Conclusion will be presented.

Please note that I have presented analysis of the original "Abstract" which is the first edition. However, I have examined the last edition, the sixth, and compared it to the first. Little has been changed, whether in arguments or conclusions. Even the Introduction is practically the same, including apologies for lack of data and not including references, except for updating verbal tense and phrases to the present.

EXAMPLE:

Last paragraph first edition: "the view which most naturalists entertain,"

Last paragraph sixth edition: "the view which most naturalists until recently entertained,"

Although the full text of the two editions was only minimally compared, the analysis was thorough, word for word, of the Table of Contents, Introduction, Headers of each chapter and the Recapitulation & Conclusion. None of the changes appear to change the overall meaning of the work or add meaningful data; they only fortify or restate the earlier version with some added examples and some references to their sources.

In the sixth edition, the Historical Sketch and the quotations following the Table of Contents are new additions to the work as is Chapter VII. "Miscellaneous Objections to the Theory of Natural Selection." Critique of these will follow the analysis of the first edition sections below.

INTRODUCTION

Paragraph 1.

"When on board the H. M. S. 'Beagle,' as naturalist, I was much struck with certain facts in the distribution of the inhabitants of South America, and in the geological relations of the present to the past inhabitants of that continent. These facts seemed to me to throw some light on the origin of species – that mystery of mysteries, as it has been called by one of our greatest philosophers. On my return home, it occurred to me, in 1837 that something might perhaps be made out on the question by patiently accumulating and reflecting on all sorts of facts which could possibly have any bearing on it. After five years' work I allowed myself to speculate on the subject, and drew up some short notes; these I enlarged in 1844 into a sketch of the conclusions, which then seemed to me probable: from that period to the present day I have steadily pursued the same object. I hope that I may be excused for entering on these personal details, as I give them to show that I have not been hasty in coming to a decision."

Paragraph 1 analysis:

These "personal details" appear to be given to establish primacy over Wallace. The dates given for the original idea (1837) and the sketch (1844) as well as assertions of working steadily to the present, (1859), establish a pattern of work over years. The sketch, included in a letter to a friend, was a simple branching tree of possible connections between similar unnamed species.

Paragraph 2.

"My work is now nearly finished; but as it will take me two or three more years to complete it, and as my health is far from strong, I have been urged to publish this Abstract. I have more especially been induced to do this, as Mr. Wallace, who is now studying the natural history of the Malay Archipelago, has arrived at almost exactly the same general conclusions that I have on the origin of species. Last year he sent to me a memoir on the subject, with a request that I would forward it to

Sir Charles Lyell, who sent it to the Linnaean Society, and it is published in the third volume of the Journal of that Society. Sir C. Lyell and Dr. Hooker, who both knew of my work – the latter having read my sketch of 1844 - honoured me by thinking it advisable to publish, with Mr. Wallace's excellent memoir, some brief extracts from my manuscripts."

Paragraph 2 analysis:

This paragraph attempts to explain why he is publishing now, although he asserts his work is incomplete, and establishes Wallace as second to arrive at nearly the same conclusion. It also provides a witness to the 1844 sketch and two witnesses to his work in general.

Paragraph 3.

"This Abstract, which I now publish, must necessarily be imperfect. I cannot here give references and authorities for my several statements; and I must trust to the reader reposing some confidence in my accuracy. No doubt errors will have crept in, though I hope I have always been cautious in trusting to good authorities alone. I can here give only the general conclusions at which I have arrived, with a few facts in illustration, but which, I hope, in most cases will suffice. No one can feel more sensible than I do of the necessity of hereafter publishing in detail all the facts, with references, on which my conclusions have been grounded; and I hope in a future work to do this. For I am well aware that scarcely a single point is discussed in this volume on which facts cannot be adduced, often apparently leading to conclusions directly opposite to those at which I have arrived. A fair result can be obtained only by fully stating and balancing the facts and arguments on both sides of each question; and this cannot possibly be here done."

Paragraph 3 analysis:

Darwin seems to be saying: "Believe me, please. I'm not including any data or references to back me up, but you should believe me anyway." This is a curious, convoluted explanation of why no references or attributions are given and also absolves him of any glaring errors the reader may find – as an incomplete, rather than a hastily written, work. First, he states that the work is not complete and lacks references and facts, and then he explains that reaching his conclusions, rather than the opposite view, requires complete facts. What is also puzzling to me is how anyone does research without including every single reference and attribution with the data as it is collected or documented. Without attribution, any unique information from another source is a form of plagiarism or theft of intellectual property.

Those who argue that Darwin delayed revelation of his theory because of fear of rejection and loss of scientific reputation would probably argue that he was afraid to include the names of other scientists or works lest their scientific reputations be tarnished by association. There is no implicit or explicit hint of such thought processes here. Remember that Darwin's circle freely discussed other evolutionary theories to scientifically explain the diversity of life on earth without including God as a reason. It was fashionable among disciples of the Enlightenment to assume science was of a higher order if consideration of God were excluded. It is interesting to note that the sixth edition, (and I assume the intervening editions), does not include any more concrete observational or experimental facts or significant additional references and attributions.

Paragraph 4.

"I much regret that want of space prevents my having the satisfaction of acknowledging the generous assistance which I have received from very many naturalists, some of them personally unknown to me. I cannot, however, let this opportunity pass without expressing my deep obligations to Dr. Hooker, who for the last fifteen years has aided me in every possible way by his large stores of knowledge and his excellent judgment."

Paragraph 4 analysis:

Here he claims assistance from many naturalists without naming any but his friend Dr. Hooker, who is also named as one of the two witnesses to his work. This is very curious as I have stated in the analysis of the previous paragraph. This attempts to establish him and his theory as being in the mainstream of science of the day without actually mentioning any other sources. Note that throughout the text he gives various examples but no theories with attributions.

Paragraph 5.

"In considering the Origin of Species, it is quite conceivable that a naturalist, reflecting on the mutual affinities of organic beings, on their embryological relations, their geographical distribution, geological succession, and other such facts, might come to the conclusion that each species had not been independently created, but had descended, like varieties, from other species. Nevertheless, such a conclusion, even if well founded, would be unsatisfactory, until it could be shown how the innumerable species inhabiting this world have been modified, so as to acquire that perfection of structure and coadaptation which most justly excites our admiration. Naturalists continually refer to external conditions, such as climate, food, et cetera, as the only possible cause of variation. In one very limited sense, as we shall hereafter see, this may be true; but it is preposterous to attribute to mere external condition, the structure, for instance, of the woodpecker, with its feet, tail, beak, and tongue, so admirably adapted to catch insects under the bark of trees. In the case of the mistletoe, which draws its nourishment from certain trees, which has seeds that must be transported by certain birds, and which has flowers with separate sexes absolutely requiring the agency of certain insects to bring pollen from one flower to the other, it is equally preposterous to account for the structure of this parasite, with it relations to several distinct organic beings, by the effects of external conditions, or of habit, or of the volition of the plant itself."

Paragraph 5 analysis:

He uses two examples to argue for "mutual affinities" and inter-dependence of species as a reason to believe in gradual change of species, not just adaptation to the environment, not inherited acquired characteristics, not independently created, but descended with modification like varieties.

Paragraph 6.

"The author of the 'Vestiges of Creation' would, I presume, say that, after a certain unknown number of generations, some bird had given birth to a woodpecker, and some plant to the mistletoe, and that these had been produced perfect as we now see them; but this assumption seems to me to be no explanation, for it leaves the case of the coadaptation of organic beings to each other and to their physical conditions of life, untouched and unexplained."

Paragraph 6 analysis:

'Vestiges of Creation' by Robert Chambers, 1844, advocated evolution by acquired characteristics with sudden change from one species to another, complete with adaptations, also known as saltation. Steven J. Gould and Niles Eldredge, in 1972, offered a form of this as punctuated evolution, whereby long periods of stable species were interrupted by rapid or sudden change. Darwin here rightly argues that coadaptation cannot be explained by such means. What he doesn't say is that evolution by gradual modification and natural selection cannot really explain co-adaptation and inter-dependence either.

Paragraph 7.

"It is, therefore, of the highest importance to gain a clear insight into the means of modification and coadaptation. At the commencement of my observations it seemed to me probable that a careful study of domesticated animals and of cultivated plants would offer the best chance of making out this obscure problem. Nor have I been disappointed; in this and in all other perplexing cases I have invariably found that our knowledge, imperfect though it be, of variation under domestication afforded the best and safest clue. I may venture to express my conviction of the high value of such studies, although they have been very commonly neglected by naturalists."

Paragraph 7 analysis:

Directed modification in domestication forming varieties is used as a tool to understand random, undirected modification forming new species. However, the range of modification of varieties was and is well known to be limited about a norm; none produce new species or even significantly new functions within varieties.

Paragraph 8.

"From these considerations, I shall devote the first chapter to this Abstract to Variation under Domestication. We shall thus see that a large amount of hereditary modification is at least possible; and, what is equally or more important, we shall see how great is the power of man in accumulating by his Selection successive slight variations. I will then pass on to the variability of species in a state of nature; but I shall, unfortunately, be compelled to treat this subject far too briefly, as it can be treated properly only by giving long catalogues of facts. We shall, however, be enabled to discuss what circumstances are most favourable to variation. In the next chapter the Struggle for Existence amongst all organic beings throughout the world, which inevitably follows from their high geometrical powers of increase, will be treated of. This is the doctrine of Malthus, applied to the whole animal and vegetable kingdoms. As many more individuals of each species are born than can possibly survive; and as, consequently, there is a frequently recurring struggle for existence, it follows that any being, if it vary however slightly in any manner profitable to itself, under the complex and sometimes varying conditions of life, will have a better chance of surviving, and thus be *naturally selected*. From the strong principle of inheritance, any selected variety will tend to propagate its new and modified form."

Paragraph 8 analysis:

Outlining the first three chapters, Darwin claims man-directed breeding (Chapter 1) can result in wide variation through successive selection within species and thus concludes that new species can be formed by a similar process in nature (Chapter

2) and be selected through the struggle for existence (Chapter 3). Accumulation of variations that give even a slight advantage over others would result in a better chance to survive. This is quite a leap because there is no direct evidence for wide variation within species much less the creation of new species. He admits that his theory is based on the economic theory of Thomas Malthus in *An Essay on the Principles of Population*, 1798, which predicted starvation because populations were increasing exponentially while food supplies were increasing arithmetically. Both theories rely on scarcity of resources and competition. What Darwin never fully appreciated was that any overabundance of offspring (including eggs, pollen, seed) provides food for other animals, enrichment of soils for vegetation and even sustenance for microscopic creatures such as bacteria, algae, plankton and fungi, which in turn are eaten by larger creatures. In truth, excesses are necessary for the entire system to operate. In nature, nothing goes to waste. Everything both eats and is ultimately eaten. The overabundance is necessary to insure that some survive so that each species continues.

Paragraph 9.

"This fundamental subject of Natural Selection will be treated at some length in the fourth chapter; and we shall then see how Natural Selection almost inevitably causes much Extinction of the less improved forms of life and induces what I have called Divergence of Character. In the next chapter I shall discuss the complex and little known laws of variation and of correlation of growth. In the four succeeding chapters, the most apparent and gravest difficulties on the theory will be given: namely, first, the difficulties of transitions, or in understanding how a simple being or a simple organ can be changed and perfected into a highly developed being or elaborately constructed organ; secondly, the subject of Instinct, or the mental powers of animals; thirdly, Hybridism, or the infertility of species and the fertility of varieties when intercrossed; and fourthly, the imperfection of the Geological Record. In the next chapter I shall consider the geological succession of organic beings throughout time; in the eleventh and twelfth, their geographical distribution throughout space; in the thirteenth, their classification or mutual affinities, both when mature and in an embryonic condition. In the last chapter I shall give a brief recapitulation of the whole work, and a few concluding remarks."

Paragraph 9 analysis:

In this paragraph, which is practically self-explanatory, he outlines the contents of chapters 4 through 14. In chapter 4 and 5 he discusses natural selection, which he claims to cause extinction, and basic laws of variation and correlation of growth. In chapters 6 through 9 he outlines the four difficulties he sees to his theory: 1. Transitions between species; 2. Instinct and animal mental powers; 3. Hybridism and the infertility between species; and 4. Sketchy geological record. In chapters 10 through 13 he discusses the distribution of animals throughout time and space (geologically and geographically) and their classification. Chapter 14 is a summary and conclusions.

Paragraph 10.

"No one ought to feel surprise at much remaining as yet unexplained in regard to the origin of species and varieties, if he makes due allowance for our profound ignorance in regard to the mutual relations of all the beings which live around us. Who can explain why one species ranges widely and is very numerous, and why another allied species has a narrow range and is rare? Yet these relations are of the highest importance, for they determine the present welfare, and, as I believe, the future success and modification of every inhabitant of this world. Still less do we know of the mutual relations of the innumerable inhabitants of the world during the many past geological epochs in its history. Although much remains obscure, and will long remain obscure, I can entertain no doubt, after the most deliberate study and dispassionate judgment of which I am capable, that the view which most naturalists entertain, and which I formerly entertained – namely, that each species has been independently created – is erroneous. I am fully convinced that species are not immutable; but that those belonging to what are called the same genera are lineal descendants of some other and generally extinct species, in the same manner as the acknowledged varieties of any one species are the descendants of that species. Furthermore, I am convinced that Natural Selection has been the main but not exclusive means of modification."

Paragraph 10 analysis:

Here he decries our limited knowledge of interrelationships between species at present and throughout history. He emphatically states his belief that species are not immutable, but change through time, that all living species are modified descendents of similar extinct species and that natural selection is the main means of these changes. What he fails to say is that the general belief of the day was not one of separate creations, but rather some form of evolution and interconnectedness. By oversimplifying he presents a straw man argument that can be said to be easily defeated.

ON THE ORIGIN OF SPECIES,
CHAPTER XIV. RECAPITULATION AND CONCLUSION (CRITIQUE)

Recapitulation:

1. He *assumes* that his point has been proven, regardless of missing data and links. ("Don't question me. I'm right.")
2. He *assumes* that all objections are without merit. ("Don't question me. I'm right.")
3. He *assumes* "... A struggle for existence leading to the preservation of each profitable variation." This belief is based on Malthus' economic theories.
4. He *assumes* that "... the truth of these propositions cannot, I think, be disputed." ("Don't question me. I'm right.")
5. He *believes* the sterility of worker ants is a difficulty for the theory. This is a problem for him because he wrongly *assumes* that habits are inherited as instincts, aka acquired characteristics, and sterile worker ants can't pass on their accumulated habits. How does the queen pass on instincts when she has not accumulated the habits?
6. He *assumes* that infertility of hybrids is due to physical differences that block breeding. We now know that genetic differences are the main reasons for infertility of hybrids.
7. He wrongly *assumes* that domestication removes interspecies infertility; based on fertility of varieties, he *assumes* that it points to fertility between species. This is one of the core weaknesses of the theory, because limits to variability have been consistently confirmed by experience, both before and since Darwin's time.
8. He uses domestic (guided) variety crossings to make the leap in reasoning to (*belief* in) unguided species crossings. Domestic breeding has intelligence behind it and is only between varieties of the same species, but the latter is assumed to have no limits and no guidance.
9. He *assumes* that species of the same genus or higher order must have descended from a common, now extinct, ancestor. While there may be some merit to this, he gives no data to support it, only *speculation* based on similarities.
10. He *assumes* that similar species that are separated geographically have traveled from their original area, facilitated by long periods of time. He admits he does not know how, but *speculates* that glacial periods may have helped.

11. He sees a problem in not finding gradual gradations between species.

12. He *assumes* we will not find direct links between living species, but *assumes* links will be found between living and extinct species.

13. He *assumes* that the absence of direct links is due to the slowness of changes and that few are changing at any one time.

14. He *assumes* that we don't find a series of forms in the geological strata because our *knowledge is incomplete.* "... falsely appear to have come in suddenly on the several geological stages?" It's not possible to prove a negative or prove a point based on absence of evidence.

15. He *assumes* that such forms will be found in the future but, after 150 years of research, that has not happened. ("... such strata must somewhere have been deposited...")

16. He covers any future failure to find intermediate steps by *surmising* that the geologic record *may be* very incomplete, but doesn't give any reasons why he *thinks* so. "I can answer these questions and grave objections only on the *supposition* that the geological record is far more imperfect than most geologists believe." It is interesting that he again sees absence of evidence as evidence itself.

17. He *assumes* that there has been sufficient time for the changes he *assumes*.

18. He *assumes* that we will not recognize links unless we find the intermediate stages, but that we *should expect* to find those missing stages in the future

19. He *assumes* that intermediate forms, if found, will be seen as separate species and not certain links, and that intermediates *may be* varieties not really separate species.

20. He *assumes* that "Local varieties will not spread into other distant regions until they are considerably modified and improved;" But gives no reason why this is *assumed* to be true.

21. He *assumes* that intermediates, "... when they do spread, if discovered in the geological formation, will appear as if suddenly created there, and will be simply classed as new species." In other words, he *assumes* that intermediates *may not be* recognized as such.

22. He sees (*believes*) the sudden appearance of many animal types in the lowest fossil bearing strata, the so-called Cambrian[1] Explosion, as a problem for his theory. He explains the *absence of evidence of earlier life*, referring to Chapter IX, by *speculating* that before that time the continents were ocean floor and the oceans were continents by a

popularly held theory at the time of oscillation between subsidence and elevation; the continents thus would only accumulate fossils before elevation and earlier fossils must be hidden under the open oceans. Presently, there is *no evidence* of this worldwide oscillatory process. Additionally, the ocean floor differs from continents in its geology, ie, heavy basaltic rock vs. lighter continental rock.

23. He *assumes* that his theory of Decent with Modification is valid and *claims* that objections to it are based, not on facts, but on ignorance about the many intermediate forms, the imperfect geological record and imperfect knowledge of means of distribution geographically. ("Don't question me. I'm right.")

24. He *assumes* that domestication reveals and concentrates natural variations and that Struggle for Existence, given scarce resources, is powerful enough to explain formation of new species over time.

25. He *dismisses* the limits to variability as an unproven assertion "It has often been asserted, but the assertion is quite incapable of proof, that the amount of variation under nature is a strictly limited quantity." The limits to variability within a species have been well recognized through centuries of experimentation. An organism can only vary within a given window and a return to original type is not uncommon. No new species have ever been produced by domestic breeding.

26. He *assumes* that species are just "strongly marked and permanent varieties" and that all species first existed as varieties; he *believes* there is little or no difference between species and varieties. A major flaw of the theory is that he *assumes* that the limits to variability do not exist.

27. He cites examples of what he *believes* are imperfections in nature as arguments against independent creations and immutability of species, and as support for his theory of Descent with Modification. He *assumes* that these "imperfections," eg, bee's sting causing its death, drones killed after mating, excessive pollen production are intermediates on their way to perfection. It was a *popular belief* of the time that the world, both physically and socially, is naturally progressive and moving toward perfection.

28. He *assumes* that species well adapted to their environment have been modified through use or disuse of capabilities over time. So blind cave creatures are *assumed* to have evolved through disuse of their vision. This is just one step away from the discredited Lamarckian theory of acquired characteristics. It is based on Darwin's *belief* in

pangenesis, whereby gemmules are shed by all body cells and migrate to the reproductive cells (ova and sperm). Parts that are disused would atrophy, and this acquired characteristic would be inheritable through the action of the gemmules. This *belief* was a great embarrassment to modern evolutionists and brought about Neo-Darwinism in which random mutation of DNA, not gemmules, is recognized as the means of inheritance, and use and disuse are down-played or ignored.

29. He *assumes* there is a correlation of growth so that when one part is modified, other parts are necessarily also modified to fit, but he gives no reason why this should be so.

30. He *assumes* that reversion to original type or appearance of a less developed part in varieties and species is evidence of his theory. What it really shows is the well-known limitations to variability.

31. He *assumes* that species vary more in the characteristics that have been more variable in the past, than in characteristics that have not changed for a long period of time – as if long endurance somehow makes them permanent or at least less apt to change. In other words, varieties and species are *assumed* to have characteristics that vary at different rates.

32. He *assumes* that instinct varies in the same way as physical characteristics and can be affected by habits. If habit can be passed on as modified instinct, he *believes* there is a problem in passing on the habits of worker ants that are sterile. Again, this is a form of acquired characteristics from environment, use and disuse, and habit.

33. He *assumes* that instincts that, by his definition, are not perfect or prone to errors or are detrimental to the animal support his theory as opposed to separate creations, implying that a creator would not create imperfect creatures. This *assumes* that the creatures in question really are flawed, not different in ways he does not understand, as if he can know the mind of God.

34. He *assumes* that the geologic record, being extremely imperfect, supports his theory. Again, absence of evidence is seen as evidence itself.

35. He *assumes* parallel forms, widely separated geographically, have migrated and then have been modified similarly due to the same means of modification, ie, the environment, existing for both.

36. He *assumes* that similar environments with different species have been separated for a long time, and would have been colonized by different species and therefore will have been modified in different ways.

37. He *assumes* that barriers to migration account for presence of species that could pass the barrier and absence of species that could not. eg, bats on islands, but not other mammals.

38. He *assumes* that species on islands that are similar to their mainland counterparts is evidence against special creations.

39. He *assumes* that similar bone structures between vertebrate species supports a common ancestor and his theory of modification and survival of species.

40. He *assumes* that embryological similarities and later divergence between mammals, fish, et cetera, supports his theory.

41. He *assumes* disuse of parts will reduce them, and that rudimentary organs in the young may be lost in mature organisms if they are not useful.

ANALYSIS OF CONCLUSIONS OF
ORIGIN OF SPECIES, 1ST EDITION

Note that many of the conclusions are redundant of recapitulations above.

Darwin asks: "Why, it may be asked, have all the most eminent living naturalists and geologists rejected this view of the mutability of species?" He then answers it. See the points listed below and my comments on each point where needed.

1. "It cannot be asserted that organic beings in a state of nature are subject to no variation;..."

    False argument or straw man. Varieties in nature were well recognized. No variation is not what others asserted, but rather limited variation within a species as seen in breeding (and has been confirmed to the present day). Naturalists of the day also recognized that there had been species changes in the past

2. "... it cannot be proved that the amount of variation in the course of long ages is a limited quantity;..."

    Proving a negative. In spite of the known observations and facts, absolute proof is impossible because we can't go back in time and observe every incident through the ages. Although it has never been observed to the present day, he argues that it *could* have happened. After breeding thousands of generations of short lived species such as fruit flies and bacteria, no new species have emerged.

3. "... no clear distinction has been, or can be, drawn between species and well-marked varieties."

   This is a *false statement*. By definition, species crossing is either impossible or, when facilitated artificially, results in sterile offspring; varieties can cross and offspring are fertile. Note that there have been recent attempts to blur the lines between species and varieties. Species, by this misguided scheme, may only mean varieties that are separated geographically, so that crossing in nature is impossible. This serves the agenda of evolutionists and environmentalists so that they can declare these "unique species" as endangered although others of their species exist elsewhere in abundance. It also advances the theory that species are just well defined varieties and that variability is not limited over time.

4. "It cannot be maintained that species when intercrossed are invariably sterile, and varieties invariably fertile."

   Proving a negative. On the other hand, if species and varieties are interchangeable by his definition, then crossed species might be fertile because they are really varieties.

5. "…we are apt to assume, without proof, that the geological record is so perfect that it would have afforded us plain evidence of the mutation of species…"

   False argument or straw man. The real objection is not that the record is expected to be perfect, but that there is no evidence so far of any intermediate steps and to the present there is none in spite of a few claims to the contrary.

6. "... we are always slow in admitting any great change of which we do not see the intermediate steps."

   While this is certainly true, it says nothing about the validity of the new ideas. Skepticism is a necessary part of testing any scientific (or other) theory.

7. "The mind ... cannot add up and perceive the full effect of many slight variations, accumulated during an almost infinite number of generations."

   Specious argument. This is based on his mistaken opinion that others may not be smart enough to surmise gradual change over long ages.

8. "… hide our ignorance under such expressions as the 'plan of creation,' 'unity of design,' et cetera, and to think we give an explanation when we only restate a fact."

   The Theory of Evolution also falls into this category. It is just a statement of inferred conclusions from observations and does not really explain anything.

9. "... attach more weight to unexplained difficulties …"

   He is implying that the difficulties are unimportant to the validity of the theory. Ignoring "unexplained difficulties" does not make a theory more valid.

10. "Future acceptance of this theory is left to the younger scientists who are encouraged to express their belief in it."

    This is a political, rather than a scientific statement.

11. "Some scientists have questioned whether some 'reputed species in each genus are not real species; but that other species are real'."

    He wants to blur the lines between the two. Varieties may be mistaken for separate species if they are very diverse and have not been observed to intercross with viable, fertile offspring.

12. He then goes on to argue whether anyone can tell which are the "special creations" and which are from "secondary causes."

    This is obviously a derogatory, almost mocking, reference to some who said that God created everything but has used secondary causes to modify them to account for the diversity we find.

13. "But do they really believe that at innumerable periods in Earth's history certain elemental atoms have been commanded suddenly to flash into living tissues?"

    This is a *False argument* against successive creations, from atoms, rather than guided change as meant by most naturalists of the day.

14. "Organs in a rudimentary condition plainly imply that an early progenitor had the organ in a fully developed state;"

    I'm not sure why he *assumed* that this was a remnant of an earlier form and not one on its way to being formed into a complete organ. Both of which are *assumptions* without supporting data.

15. "I *believe* that animals have descended from at most only four or five progenitors, and plants from an equal or lesser number. …all living

things have much in common ... probably all the organic beings which have ever lived on the earth have descended from some one primordial form, into which Life was first breathed."

This is an opinion based on a *belief*, not science.

16. Prediction: "... there will be a considerable revolution in natural history." And end disputes about which are true species and which are varieties.

    This is a self-evident statement of a *belief and hope*.

17. He *claims* that the only difference between species and varieties is that varieties have living intermediates whereas species had them in the distant past.

    Here again, he *assumes* that species and varieties are practically interchangeable, only on a different time scale.

NOTE: the following predictions are self-evident statements of beliefs and hopes.

18. Prediction: classification terms such as affinity, relationship, et cetera, would cease being metaphors and become actual connections.

    > Classification uses these terms to describe similarities in the form and functions of species so that they fall into families, genera, et cetera, based on these similarities. He *believes* that his theory will result in these similarities being defined as actual generational connections.

19. Prediction: New fields of inquiry "... on the Laws of variations, on correlation of growth, on the effects of use and disuse, on direct action of external conditions, and so forth."

    > Such studies have only produced negative results. Instead we have Neo-Darwinism which emphasizes genetic variability.

20. Prediction: the study of domestic productions will rise in importance.
21. Prediction: Classifications will become genealogies and rules will be simplified.
22. Prediction: Living fossils will help clarify ancient forms.
23. Prediction: Embryology will reveal progenitors.
24. Prediction: Increased knowledge will trace migrations and determine the means.

25. He *assumes* that the fossil record is very incomplete, that vast ages pass between the rare fossil formations, and predicts that the duration of the gaps will be determined by the amount of difference between the fossils.

> Here he *assumes* a specified rate of change, although he earlier states that different parts are modified at different rates. Modern work also uses this same logic by judging the amount of difference between genes, but the genes for different characteristics are known to vary at different rates.

26. He *assumes* that species are formed by slow acting and still existent conditions, not God or catastrophes.

> This is *speculation* based on the popular belief of time in the uniformitarian theory, by which it is assumed that conditions on earth have not changed dramatically through the ages. He *believed* that long, uninterrupted ages were necessary for his theory to work and that any interruption in the vast ages could jeopardize his theory.

27. He *assumes* that improvement of one species will result in improvement or extinction of others and that little or no change may happen over long periods.

> This is a statement of *belief,* without any evidence given, to explain succession of species through geologic time.

28. He *assumes* that early forms of life were fewer and simpler, that they evolved slower and that the earliest evolved extremely slowly.

> This is a statement of *belief,* but he gives no reason why he believes that earlier forms evolved more slowly – perhaps he assumed there was less competition.

29. Prediction: Psychology will study the "acquirement of each mental power and capacity by gradation." including the "origin of man and his history."

> This is a self-evident statement of a *belief and hope.*

30. He *assumes* that species formation by secondary causes, not separate creations is somehow superior.

> This is an opinion based on a *belief,* not science.

31. He *assumes* that no living creature will be unchanged in the future, and that few will have descendants and will become extinct.

    This is an opinion based on a *belief*, not science. He doesn't explain panchronic species.

32. He *assumes* that all creatures have descended from a few progenitors and that the line has never been broken by a worldwide cataclysm.

    This is an opinion based on a *belief*, not science.

33. He *assumes* that evolution is working progressively toward perfection.

    This is an opinion based on a *belief*, not science.

34. He *assumes* variability is caused by "… direct and indirect action of the external conditions of life, and from use and disuse."

    This is an opinion based on a *belief*, not science.

35. He *assumes* that overproduction ("Ratio of Increase") will continue to result in a "struggle for life and to natural selection… divergence of character, the extinction of less-improved forms."

    This is an opinion based on a *belief*, not science.

MY CONCLUSIONS:

This is a series of conjectures without foundation except for Darwin's faith in his own prejudices and assumptions. My conclusion is that Darwin's work is little more than philosophy, speculation and pure guesswork, some of which are clearly wrong. It is one long opinion piece. He offers no indisputable supporting facts other than a series of assumptions based on similarity of form and a projection of his work with domestic breeding onto unguided nature, based on Malthus' economic philosophy of population limitation through food scarcity. True logical, scientific connections are at best tenuous, and at worst lacking. While this work by Darwin is an excellent example of nineteenth-century descriptive biology, as a modern scientific work it does not pass the smell test. It would never pass peer review by today's standards. This is not to say there is not some merit in such speculations, but that it is not science. It is philosophy.

His whole argument hangs on a few wrong or unsupported assumptions such as:

- unlimited variation over time into new species;

- inherited acquired characteristics, whether physical from environmental influences, use or disuse, or instinct and abilities from habit;

- similar forms imply a common ancestor; fossils as ancestors of living species; gradual steps not found due to incompleteness of geological record;

- sudden appearance in the Cambrian explained as earlier life now under seas and incomplete geological record;

- embryological similarities or similarities of form indicate a common ancestor; correlation of growth.

Difficulties he identified include:

- absence of intermediates;

- incomplete fossil record;

- means of migration of widely separated species;

- inheritance of sterile worker habits in colony insects;

- sterility of hybrids.

I have read your book with more pain than pleasure. Parts of it I admired greatly; parts I laughed at till my sides were almost sore; other parts I read with absolute sorrow; because I think them utterly false & grievously mischievous — You have deserted—after a start in that tram-road of all solid physical truth—the true method of induction—& started up a machinery as wild I think as Bishop Wilkin's locomotive that was to sail with us to the Moon. Many of your wide conclusions are based upon assumptions which can neither be proved nor disproved. Why then express them in the language & arrangements of philosophical induction?

Adam Sedgewick, noted geologist who taught Darwin,
after reading Origin of Species

APPENDIX A, PART 2

SIXTH EDITION CRITIQUE

Analysis of portions of On the Origin of Species by Means of Natural Selection or The Preservation of Favoured Races in the Struggle for Life, sixth edition, London, 1872

Analysis of: "An Historical Sketch of the Progress of Opinion on the Origin of Species, Previously to the Publication of the First Edition of the Work"

Starting with Aristotle, Darwin here gives a review of the history of theories about the mutability of species. He acknowledges many others whose works have promoted Evolutionary views both before and after him as follows: Aristotle, Jean-Baptiste Lamarck, Buffon, Geoffery Sainte-Hilare, Erasmus Darwin, Goethe, W. C. Wells, Rev. W. Herbert, Robert Grant, Patrick Matthew, Von Buch, Rafinesque, Haldeman, Robert Chambers, M. J. d'Omalius d'Halloy, Dr. Freke, Herbert Spencer, M. Naudin, Unger, Dalton, Oken, Gordon, Count Keyserling, Dr. Schaaffhausen, M. Lecoq, Rev. Baden Powell, Sir John Herschel, Darwin and Wallace, Von Baer, Huxley, Hooker. It is obvious in the text that many readers of his book have contacted him about others who had similar theories before him. Note that those people included differ from my list gleaned from modern sources. All of the comments below are condensations of Darwin's opinions.

Aristotle – "Physicae Auscutationes" - believed that the parts of the body were assembled by chance and that the parts that proved useful were preserved

Erasmus Darwin (his Grandfather) – 1794 "Zoonomia." He anticipated Lamarck in inheritance of acquired characteristics.

Goethe – 1794 believed in species changing and posed a question for naturalists to determine how characteristics were acquired.

Geoffery Saint-Hilaire – 1795 "Hist. Nat. Generale" mentions Buffon's earlier work. Conditions of life cause change and species were originally varieties. A few original plastic forms changed but that change is not happening in the present.

Jean-Baptiste Lamarck – 1801, 1809 "Philosophie Zoologique"; 1815 "Hist. Nat. des Aminaux san Vertebres." (Natural History of Invertebrate Animals) – He believed that all species have changed through acquired characteristics, that all change in nature occurs through nature's laws, that life is progressive toward higher forms, that inheritance is through direct action of physical conditions, crossings, use and disuse and habits. He believed that the existence of simple forms in the present could be explained by on-going spontaneous generation.

Darwin credits him with originating the thought that all creatures have descended from one or a few ancestors, and that all change takes place through laws of nature. He also mentions that his theory of change by acquired characteristics has been discredited. I'm not sure how this differs from Darwin's change through use and disuse, habit, and inheritance by gemmules of a parent's circumstantially derived characteristics, but he definitely is trying to distance himself from inheritance of acquired characteristics.

W. C. Wells – 1813 – "An Account of a White Female, Part of Whose Skin Resembles that of a Negro." Animals change, breeders improve animals through selective breeding and nature could do the same, only slower. Evolution of man's races is emphasized as resulting from environmental conditions and a linked resistance to certain diseases.

Rev. W. Herbert – 1822 – "Horticultural Transactions" and 1837 "Amaryllidaceae" Botanical species are only higher, more permanent varieties and this extends to animals. A few original plastic species made all the others through crossings and natural variation.

Robert Grant – 1826 – paper in "Edinburgh Philosophical Journal" Species have descended from other species and life is progressive.

Patrick Matthew – 1831 – "Naval Timber and Arboriculture" Darwin acknowledges that Matthew had the same theory as his at an earlier time but that it was overlooked because it was hidden in this larger work.

Von Buch – 1836 – "Description Physique des Isles Canaries" Varieties slowly change into species.

Rafinesque – 1836 – "New Flora of North America" Varieties become species except for the original ancestors.

Prof. Haldeman – 1843 – in "Boston Journal of Nat. Hist. U. States" He gives both sides of the argument of immutability but Darwin felt he leaned toward mutability.

Robert Chambers (writing anonymously) – 1844 – "Vestiges of Creation" to 10th edition in 1853. Life is naturally progressive and is modified by circumstances. Organization of creatures is sudden by leaps but other variation is slower. Darwin criticizes this popular work but credits him with opening the discussion and removing prejudices.

M. J. d'Omalius d'Halloy – 1831 and 1846 – in "Bulletins de l'Acad. Roy. Bruxelles." Species are formed by descent with modification, not separate creations.

Richard Owen – 1849 & 1858 (after Darwin/Wallace paper) – "Nature of Limbs" Archetypical ancestors become new species, not separate creations but all originated from the first Creative Cause. He claimed primacy for the theory of natural selection over Darwin and Wallace. Darwin spent some time here arguing against such claims and that the point is moot since both were preceded by both Matthew and Wells.

Dr. Freke – 1851 in "Dublin Medical Press" – all species descended from one primordial form, but Darwin claims that the means were different from his. 1861 "Origin of Species by means of Organic Affinity" seemed to confirm his opinion.

Herbert Spencer – 1852 & 1858 – He contrasted the theories of Creation vs. Development. He noted difficulties between breeding of varieties vs. species failure to produce fertile crossed offspring, changes in embryos during development and "general gradation." He also stated that species are formed due to circumstances. In 1855 he addressed development of mental powers.

M. Naudin – 1852 – Origin of Species paper in "Review Horticole" – species formed like varieties in domestication but he doesn't explain natural selection. He alleged that original ancestors were very "plastic," but later forms are final and unchanging.

Unger – 1852; Dalton – 1821; Oken; Bory St. Vincent; Burdach; Poiret; Fries also believed in mutability of species.

Count Keyserling – 1853 – believed that new diseases were caused by airborne chemicals changing "germs."

Dr. Schaaffhausen – 1853 – some species have changed; new species are due to intermediate varieties dying off.

M. Lecoq – 1854 – believed in mutability of species.

Rev. Baden Powell – 1855 – "Essays of the Unity of Worlds"
Change has happened and continues today.

Sir John Hershel – change is natural, not miraculous.

Charles Darwin and Alfred Wallace – 1858 – paper in *Journal of the Linnean Society*, vol. 3.

Von Baer – 1859 – species that are widely separated have descended from a common ancestor.

Prof. Thomas H. Huxley – 1859 - "Persistent Types of Animal Life" is a study of panchronic species as opposed to those that change.

Dr. Joseph D. Hooker – 1859 – "Introduction of the Australian Flora" includes descent with modification.

Analysis of new 6th Edition Chapter VII. Miscellaneous Objections to the Theory of Natural Selection.

See chapter title and headings from the book below.

"VII. MISCELLANEOUS OBJECTIONS TO THE THEORY OF NATURAL SELECTION. Longevity -- Modifications not necessarily simultaneous -- Modifications apparently of no direct service -- Progressive development -- Characters of small functional importance, the most constant -- Supposed incompetence of natural selection to account for the incipient stages of useful structures -- Causes which interfere with the acquisition through natural selection of useful structures -- Gradations of structure with changed functions -- Widely different organs in members of the same class, developed from one and the same source -- Reasons for disbelieving great and abrupt modifications."

Darwin, Charles (2008-02-14). *The Origin of Species by Natural Selection*, 6th edition, 1872, (Kindle Edition)

Most if not all of the Objections are from *Genesis of Species* by St. George Mivert, London, Macmillan and Company, 1871. See analysis below.

St. George Mivert. *On the Genesis of Species*, MacMillan and Company, 1871, (Kindle edition) Analysis of objections:

1. **Objection:** Disagreement with the theory that species are all imperfect.

   Answer: Species are not as perfectly suited to their environments as they could be. His evidence is that invasive species take over, but this is flawed.

   Comment: Due to lack of predators and other limiting factors, invasive species disrupt the balance in an environment, so they can't be said to be better suited for it.

2. **Objection:** Evolution should favor longer lives.

   Answer: Longer lives could be detrimental if cold periods could wipe them out. Length of life is related to the size and is compensated by greater fertility.

3. **Objection:** Species have not changed since ancient Egypt implies that none have changed throughout the world.

   Answer: Egyptian environment has not changed so no advances were needed. Only when environments change is change likely but natural selection is a random, not a necessary event.

4. **Objection:** Species differ in multiple characteristics that must have been modified at the same time.

   Answer: Darwin asserts that many slight modifications at different times and rates can account for the multiple differences seen.

5. **Objection:** Many characteristics appear to add no advantages so they can't be affected by natural selection.

   Answer: Darwin sees this as a serious objection, but asserts that we do not know whether such structures are beneficial in some way; he asserts that some parts may have been modified by correlation of growth along with other advantageous parts. He also states that whatever persistent influence caused the modification must act on all individuals of that species. This, apparently, refers to his belief that external environmental pressures cause mutations that are inheritable. He goes on to describe flowers with multiple forms on the same plants as by the "laws if growth" affected by location on the plant. He asserts that some characteristics that are not or no longer useful or that may have originally arisen from "fluctuating variations" may be passed along with other useful modifications.

6. **Objection:** Incipient stages of development cannot be accounted for by natural selection.
   Answer: Darwin gives the example of the giraffe's longer neck helping in survival, so incipient forms would ultimately die out in droughts. He again refers to inheritable use and disuse.
7. **Objection:** If the long neck is so advantageous, why have other hoofed animals not developed long necks or a proboscis to reach tree branches?
   Answer: Darwin doesn't really address this other than to say sheep do not browse on tree branches like horses or cows do, so a longer neck would not be an advantage. He also speculates that these domestic animals might develop longer necks in time if conditions favor it.
8. **Objection:** If variation occurs in all directions, how have insects come to mimic twigs, leaves, et cetera?
   Answer: Starting from an insect with some resemblance to its surroundings, those that better mimic their surroundings would survive.
9. **Objection:** In baleen whales, how is the development of baleen accomplished?
   Answer: Starting from a smaller sifting lamella like that of a duck, extended lengths can be selected over time as needed for survival. Examples of various ducks, petrels and whales are given to illustrate variation that exists.
10. **Objection:** How would gradual evolution of eyes that move to one side of the body in flat fish such as flounder be of benefit until the entire transition developed. A partial transition would be more of a handicap than a survival benefit.
    Answer: An example is given of a young fish that, due to its body shape, small lateral fins and lack of a swim bladder struggles to remain upright. It eventually sinks to the bottom and, through effort, twists the eye on the lower side to try to see. He felt that over time, this repeated effort could cause a permanent change. He also explains the changes in color and size of the lower lateral fin as due to disuse relative to the other side.
11. **Objection:** How is the prehensile tail developed and how would it provide a survival advantage before fully developed.
    Answer: Darwin cites certain mice that use their tails to both assist in climbing and to balance when jumping from twig to twig.

12. **Objection:** How could mammary glands have developed before they were fully functional?

    Answer: Darwin assumes that mammary glands on mammals were first cutaneous glands within a marsupial pouch. This looks more like a misdirection since it doesn't fully explain the origin of marsupial glands other than to say that those who developed more complete glands would offer a survival advantage to those that didn't.

13. **Objection:** How could the young "learn" to suck? How would an elongated larynx be developed in kangaroo to facilitate passage of milk and later be lost?

    Answer: Unlike mammals that suck, marsupial glands passively drip milk into the mouth of an attached fetus so the elongated larynx is necessary to facilitate breathing without strangulation.

APPENDIX B

HOW TO GET ENERGY FROM SUGAR

Glucose is the basic sugar used for energy. Other simple sugars or more complex sugars or fats must be converted to glucose before entering the process that yields CO_2, water and energy by regenerating ATP from ADP. The entire process takes place in the mitochondria within the cell. It is closely regulated by enzymes and feedback signals from products so that energy is only produced when it is needed and only in the amounts needed.

Enzymes are reaction regulators (catalysts). Most are huge, complex molecules that hold, rearrange, unfold, open or cut other molecules. The names of enzymes end in "–ase." (Sugars end in "–ose")

ATP is the basic source of energy for work within the cell. The ATP cycle is the process where ATP yields energy by giving up phosphate ions and is then regenerated by adding them back.

ENERGY PROVIDING MOLECULES

ATP – high energy source; adenosine triphosphate; each phosphate ion yields energy when removed and requires energy to add phosphate back.

ADP – adenosine diphosphate is formed by removal of one phosphate from ATP. ATP is regenerated when a phosphate is added back.

AMP – adenosine monophosphate; formed by removal of two phosphates from ATP; it is the lowest energy state. ATP is regenerated when two phosphates are added stepwise to AMP.

GTP/GDP – Guanosine Triphosphate/ Guanosine Diphosphate – a source of energy similar to ATP/ADP. It can give up a phosphate to ADP, regenerating ATP.

NAD/NADH – Nicotinamide adenine dinucleotide is a CoEnzyme facilitating certain reactions as a Hydrogen donor.

FAD/FADH – Flavin adenine dinucleotide is a CoEnzyme facilitating certain reactions as a Hydrogen donor.

Step 1 – Glycolysis turns one glucose (6 carbons) into two pyruvate (3 carbons each) and generates two extra ATP.

In a ten step process, **glucose** is cut in two and rearranged into two <u>pyruvate</u> molecules by using the energy of two ATP to ADP conversions and then generating four ATP from ADP later in the process. This process uses ten different enzymes, one for each step of the process, and two nucleotides. Three of these enzymes regulate the rate of the process, based on the level of ATP and its products ADP and AMP. The process slows down when there is high ATP and speeds up when there is high ADP or AMP and low ATP. In other words, when the energy source is low more energy is generated faster.

Step 2 – The two Pyruvate molecules lose 1 carbon each, and are attached to CoA (Coenzyme Acetylation) to form acetyl CoA

Step 3 – Citric Acid Cycle turns acetyl CoA into Energy, CO_2 and H+ with production of NADH.

Step 4 – Electron Transport Chain yields a total of 6 more ATP

Figure 1. Step 1 – Glycolysis (enzymes at each step omitted for simplicity)[1] See following table for more details.

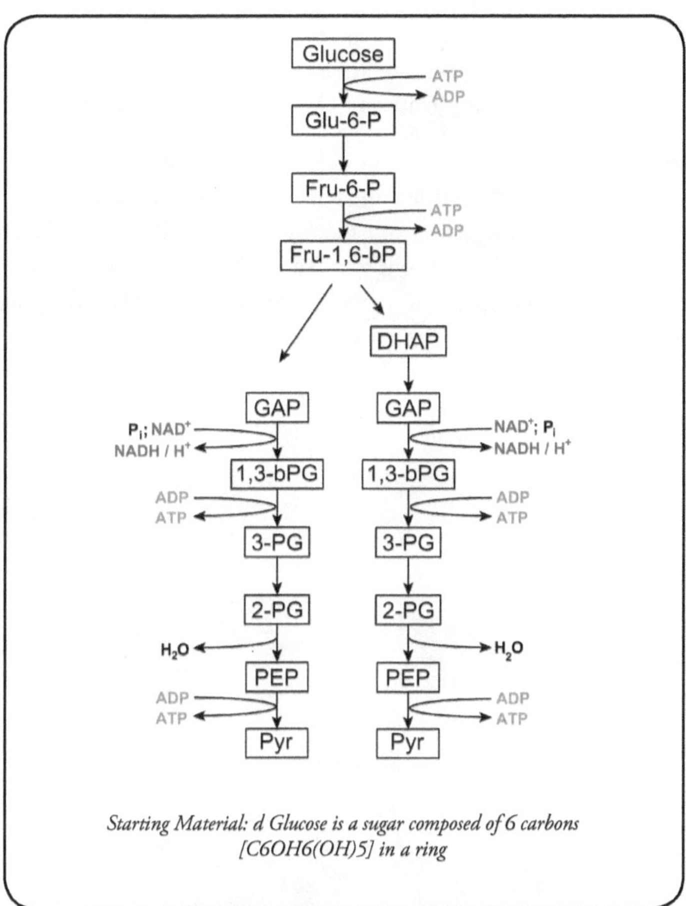

BASIC STEPS IN GLYCOLYSIS:

- Open the ring and add a phosphate: In the first step, the enzyme glucokinase, with the help of a magnesium ion, opens the ring and adds a phosphate from ATP to carbon number 6, producing D-Glucose-6-Phosphate and ADP.

- Rearrange: The enzyme phosphoglucoisomerase rearranges D-Glucose-6-Phosphate into D-Fructose-6-Phosphate.

- Add a second phosphate: The enzyme phosphofructokinase, with the help of a magnesium ion, adds a phosphate from ATP to carbon number 1 producing D-Fructose 1,6-Diphosphate

- Cut in half: The enzyme Fructose Diphosphate Aldolase cuts D-Fructose 1,6-Diphosphate to form the 3 carbon molecules Glyceraldehyde-3-Phosphate and Dihydroxyacetone Phosphate.

Since this is a very complex process with many steps and products, I struggled with how best to present it. I settled on tables as the best way to present each step as simply as possible while giving you the major details, including the enzymes at each step. Even these have left out some of the finer points, but generally give you the idea without bogging you down in complex chemical intermediates or chemistry calculations.

TABLE 1 GLYCOLYSIS PHASE 1: FROM GLUCOSE WITH 6 CARBONS TO TWO GLYCERALDEHYDE 3-PHOSPHATES WITH THREE CARBONS EACH

Step	Starting Materials	Operation	Enzyme, et cetera	Products	Comments
1#	D-Glucose and ATP	Open the ring and add Phosphate	Glucokinase, Magnesium ion (Mg++)	D-Glucose 6-Phosphate, ADP	Uses 1 energy molecule (ATP to ADP). Glucose has 6 carbons.
2*	D-Glucose 6-Phosphate	Rearrange	Phosphogluco- isomerase	D-Fructose-6-Phosphate	
3#	D-Fructose-6-Phosphate and ATP	Add a second Phosphate	Phosphofructokinase, Mg++	D-Fructose 1,6-Diphosphate, ADP	Uses 1 energy molecule (ATP to ADP) Not reversible
4*	D-Fructose 1,6-Diphosphate	Cut into two	Fructose Diphosphate Aldolase	Glyceraldehyde-3-Phosphate (GAP) and Dihydroxy-acetone phosphate (DHAP)	Splits Glucose (6 carbons) into two 3 carbon molecules. GAP can be used in the next steps but DHAP must be rearranged into GAP first.
5*	Dihydroxyacetone Phosphate	Rearrange	Triose phosphate isomerase	Glyceraldehyde-3-Phosphate	Result is 2nd GAP (3 carbons)

KEY for all Tables:

Reversible steps denoted by *

Steps governed by the available ATP/ADP denoted by #

Steps governed by the available NAD+/NADH + H+ are denoted by $

TABLE 2 GLYCOLYSIS PHASE 2:   TWO GLYCERALDEHYDE 3-PHOSPHATE MOLECULES FROM ONE GLUCOSE YIELDS DOUBLE PRODUCTS FOR THE REST OF THE PROCESS.

	Starting Materials	Operation	Enzyme, et cetera	Products	Comments
6*$	Glyceraldehyde-3-Phosphate and Nicotinamide Adenine Dinucleotide, (NAD^+) and Phosphate ion	Add a second Phosphate, generate NADH (using one H+ from the reaction)	D-Glycerladehyde phosphate dehydrogenase	3-Phosphoglyceryl phosphate, NADH+ and H^+	Two hydrogen ion carrier molecules are generated
7*#	3-Phosphoglyceryl phosphate, ADP	Regenerate ATP (remove a phosphate)	Phosphoglycerate kinase and Mg^{++}	3-Phosphoglycerate and ATP	The two energy molecules (ATP) that were used in Phase 1 are regenerated
8*	3-Phosphoglycerate	Rearrange (move phosphate from one end to middle)	Phosphoglycerase mutase and Mg^{++}	2-Phosphoglycerate	
9*	2-Phosphoglycerate	Remove water	Enolase and Potassium (K^+) and Mg^{++}	Phosphoenolpyruvate and water	
10#	Phosphoenolpyruvate and ADP	Regenerate ATP and remove phosphate	Pyruvate kinase, Mg^{++}	Pyruvate and ATP	Two extra energy molecules (ATP) are generated
	End of Glycolysis – Steps below prepare Pyruvate for storage or use in Citric Acid Cycle.				
11*$	Pyruvate and NADH+	Regenerate NAD^+	Lactate dehydrogenase	L-Lactate, NAD^+	Two hydrogen ion carrier molecules are used
12$	Pyruvate and NAD	Remove CO_2 and add to Coenzyme A	Pyruvate dehydrogenase	Acetyl CoA, CO_2, NADH	3 carbon Pyruvate becomes 2 carbon acetyl attached to CoA

273

In Glycolysis two ATP are used and four ATP are generated. The net starting materials and products formed are:

Glucose + 2 Phosphates + 2 ADP →2Lactate- + 2 H+ + 2ATP + 2 H2O

In Anaerobic (lacking oxygen) conditions, Pyruvate forms L-Lactate. Otherwise Pyruvate goes on to form CO_2 and H_2O through the Krebs or Citric Acid Cycle.

First Pyruvate must be oxidized to Acetyl and attached to CoA as Acetyl-CoA and CO_2 (release of 1 carbon from each pyruvate)

Acetyl-CoA adds Acetyl group (2 carbons) to Oxaloacetate (4 carbons) to form Citrate (6 Carbons) releasing CoA

TABLE 3: CITRIC ACID CYCLE RELEASES THE REMAINING 2 CARBONS AND GENERATES ENERGY.
NOTE: 1 OF 2 ACETYL COA IN EACH CYCLE (REPEATS)

Step	Starting Materials	Operation	Enzyme, et cetera	Products	Comments
13	Acetyl CoA and Oxaloacetate	Add Oxaloacetate	Citrate synthase	Citrate, CoA,	
14	Citrate	Rearrange (2 steps)	Aconitase, Fe^{++}	Isocitrate	
15	Isocitrate	Remove one CO_2	Isocitrate dehydrogenase and $Mg{++}$	a-Ketoglutarate and CO_2	1 carbon released as CO_2
16	a-Ketoglutarate, NAD^+ and CoA	Remove one CO_2	a-Ketoglutarate dehydrogenase complex and Mg^{++}	Succinyl-CoA and CO_2	Last carbon released as CO_2
17	Succinyl-CoA, Phosphate ion and Guanadine diphosphate (GDP)	Remove CoA and add Phosphate to GDP	Succinyl-CoA synthetase and Mg^{++}	Succinate and CoA and Guanadine Triphosphate (GTP)	Generate an energy molecule (GTP)
18	Succinate, Flavin adenine dinucleotide (E-FAD) and $2H^+$	Rearrange and add H_2 to E-FAD	Succinate dehydrogenase	Fumerate and E-FADH$_2$	FAD is bound to the enzyme (E-FAD)
19	Fumerate, H_2O	Rearrange and add H_2O	Fumerase	l-Malate	
20	L-Malate	Remove 2H+ and add to NAD+	L-Malate dehydrogenase	Oxaloacetate and NADH+ and H+	Regenerate Oxaloacetate to restart the cycle. Go to Step 13

This final step in the cycle regenerates Oxaloacetate so that the cycle can begin again with the next Acetyl CoA.

Step 3 – Citric Acid Cycle (most enzymes at each step are omitted for simplicity.) The C# boxes tell how many Carbons there are at each stage.

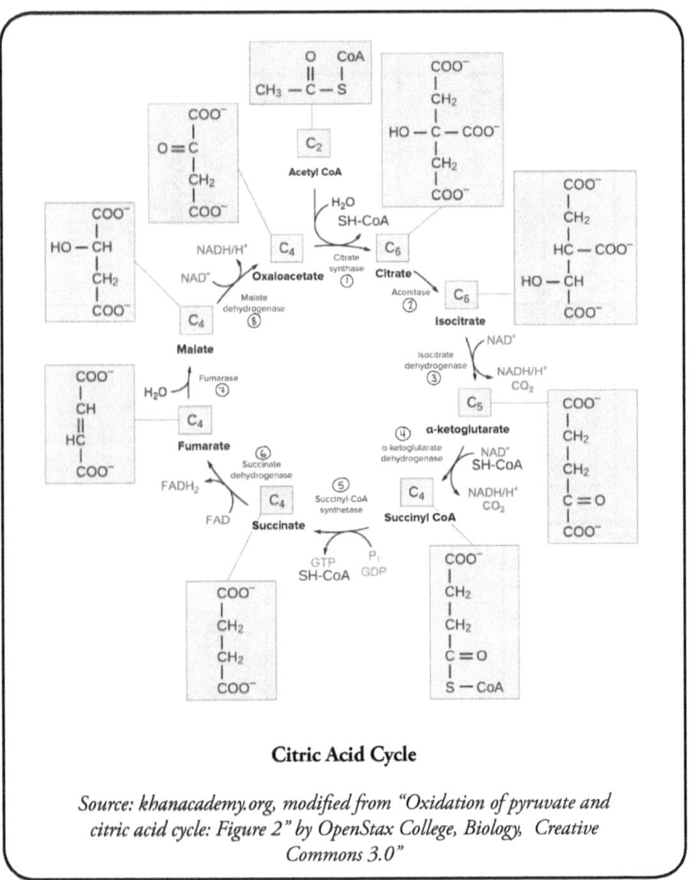

Citric Acid Cycle

Source: khanacademy.org, modified from "Oxidation of pyruvate and citric acid cycle: Figure 2" by OpenStax College, Biology, Creative Commons 3.0"

SIDE REACTIONS:

GTP generated in step 17 may give up its Phosphate to ADP to generate ATP using Nucleoside diphosphokinase enzyme.

The NADH⁺ generated in the Citric Acid Cycle enters an electron transport chain that generates 3 more ATP from ADP and 1 water for each acetyl run through the cycle – a total of 6 ATP to be used by cells to do other work.

GLOSSARY

AAAS. American Association for the Advancement of Science.

ad hoc. For this particular purpose; special; created on the spur of the moment, impromptu.

ad hominum attack. Attacking the opponent on personal grounds not related to the subject.

aether The discredited theoretical substance previously assumed to fill all of space.

amphiboly. A sentence that can be interpreted in more than one way.

analogy (biol). That which evolved independently by natural selection.

anecdote. A short narrative of an interesting, amusing or biographical incident whose reliability is not certain.

aphelion. The farthest point in a planet's orbit from the sun.

apotheosis. A perfect example; elevation to divine status.

appeal to authority. Logical fallacy whereby the views of an authority are given as an argument.

appeal to emotion. Logical fallacy: Replacing rational arguments with emotional appeals. Propaganda is often laced with emotionally charged statements.

appeal to ignorance. Logical fallacy: Because something is not known to be true, it is assumed to be false.

appeal to the majority. Logical fallacy: Consensus as proof. If most people believe it, it must be true.

appeal to motives: prejudicial language. Logical fallacy: Value or moral goodness is attached to agreeing with the argument.

a priori. From the beginning; without precedent.

argument. A statement of a conclusion supported by evidence, as opposed to assertion that offers no evidence.

argument by lack of imagination. Logical fallacy: Similar to appeal to ignorance. Drawing conclusions based on surface or obvious appearance.

argument from personal incredulity. Logical fallacy: A theory is thought to be proven untrue because someone finds it personally offensive, unlikely or unbelievable, and consequently a preferred alternative is thought to be proven true. Also known as **argument from personal belief** or **argument from personal conviction.**

argument from silence (argumentum ex silentio). Logical fallacy: Similar to appeal to ignorance; absence of evidence to the contrary or a different theory is deemed to validate the argument.

assertion. A statement of a conclusion without proper evidence.

begging the question. Logical fallacy: The premise automatically assumes the truth of the conclusion in the statements to support it.

Bell curve. The shape of a statistical data set, with fewer data points at the high and low ends.

bimodal. A statistical curve with more than one peak.

blackbody. An ideal surface that absorbs all photons. Blackbody radiation is the heat given off by such a surface that depends only on its temperature. Its spectrum is a featureless curve.

black hole. A theoretical massive cosmic body in which gravity is so great that not even light can escape.

canard. A false or unfounded report or story; a fabricated story; deceit; (Mfrench: to half-sell ducks - literally "duck").

carbon dioxide. See CO_2.

Cepheid variable star. A class of variable stars that are believed to be of the same intrinsic brightness for the same variability period, used as a standard candle in astronomy.

changing the subject. Logical fallacy: Arguing for one thing to appear to prove another.

cherry picking. The practice of selecting only the data that supports the theory or hypothesis.

clade (cladogram, cladist) A group of biological taxa (as species) that includes all descendants of one assumed common ancestor.

CO_2. carbon dioxide; a gas that is produced by animals from metabolizing (oxidizing) nutrients and from certain chemical oxidation processes such as burning fossil fuels, aka hydrocarbons. It is used by plants to produce sugars while emitting oxygen as a by-product.

complex question. Logical fallacy: The statement contains the conclusion within it.

computer model. A program made up of a set of mathematical formulae that attempts to mimic real systems in nature for the purpose of understanding or predicting outcomes.

continental drift. A theory that the continents are not fixed on the surface of the Earth and move relative to each other. Championed by Alfred Wegner in the early twentieth century. See also plate tectonics.

conundrum. A question or problem having either no answer or only a conjectural answer; a difficult or complex problem; a mystery.

cosmology. (Gk - kosmos = cosmos; logia = or reason) A branch of metaphysics that deals with the universe as an orderly system. A branch of astronomy that deals with the origin, structure and space-time relationships of the universe.

catalyst. A molecule or element that can regulate the speed and extent of a chemical reaction between other molecules without being used up in the process.

COWDUNG. COnventional Wisdom of the DomiNant Group (Michael Disney, *The Hidden Universe*, 1984).

credulous. Ready to believe, especially on slight or uncertain evidence.

dark energy. Theoretical energy opposed to gravity that has not been detected, but that may account for the expansion and possible acceleration of the universe.

dark matter. Theoretical matter that only interacts with ordinary matter gravitationally and that would account for the gravitational behavior of galaxies and may be responsible for slowing the expansion of the universe. None has been detected.

Darwinism. Belief in reductionist materialism such as evolution by natural selection and slow random change over long periods of time.

deduction. The deriving of a conclusion by reasoning, specifically: inference in which the conclusion follows necessarily from the premises.

devil's advocate. One who champions a contrary view for the sake of argument.

dialectic. Hegelian process of change in which an entity passes over into and is preserved and fulfilled by its opposite. Development through the stages of thesis, antithesis and synthesis according to the laws of dialectical materialism.

dialectical materialism. Marxist theory that maintains the material basis of a reality constantly changing in a dialectical process and the priority of matter over mind.

dogma. Established opinion put forth as authoritative, especially without adequate grounds.

Doppler effect The compression or expansion of waves (eg, light or sound) from an approaching or receding source.

dualism.
1. The theory that reality consists of two irreducible elements or modes, eg, physical and nonphysical.
2. The doctrine that the universe is under the dominion of two opposing principles one of which is good and the other evil.

ecotage. Combination of Eco (logy) and (Saba) tage. Environmental terrorism.

ecotheism. Worship of ecology (nature) to the exclusion of human considerations.

egalitarianism. A belief in human equality especially with respect to social, political, economic rights and privileges. A social philosophy advocating the removal of all inequalities among men.

emergence. The spontaneous appearance of an unexpected quality from a complex system without prior design or planning.

endothermic. A chemical reaction that absorbs more heat than is produced.

enigma. Something difficult to understand or explain.

enzyme. A large protein used in living systems as a catalyst to speed up or otherwise regulate the amount of products formed in a biochemical process. Enzymes can hold, cut, splice, rearrange or open other specific molecules to facilitate reactions without being changed themselves.

epistemology. The study or a theory of the nature and ground of knowledge especially with reference to its limits and validity.

equivocation. Logical fallacy: Use of a word or phrase to mean more than one thing; changing the meaning of a word in the middle of an argument.

ethology. The study of ethics.

eugenics. A social philosophy that deals with the improvement (as by control of human mating) of hereditary qualities of a race or breed.

evangelical. Zealously enthusiastic about promoting the spread of beliefs or worldview.

exothermic. A chemical reaction that gives off more heat than it absorbs.

exponent. The number of times minus one (n-1) that a number is multiplied by itself, written as, for example, 10^2 meaning 10 X 10 = 100.

extremophile. A term applied to microbes on earth that live in extreme conditions such as very hot or cold, very salty, acid or alkaline, low or no oxygen or where they get their energy from mineral sources instead of the sun.

fact. That which is real or demonstrated to be true.

factitious. Use of facts along with beliefs to assert untruths. Can also mean produced artificially.

fallacious. Deceptive, misleading, delusive, disappointing.

fallacy. An argument failing to satisfy the conditions of valid or correct inference.

fallacy, genetic. Logical fallacy: Holding a theory to be wrong if caused by irrational factors.

fallacy, pathetic. Logical fallacy: Ascribing human characteristics or feelings to inanimate nature.

false cause or non sequitur. Logical fallacy: Incorrectly assumes that one thing is the cause or explanation of another

falsifiable. A theory for which a case may exist that would prove the theory false. One of the necessary conditions defining scientific theories. See also verifiable.

forensics. Piecing together evidence to support a hypothesis about what probably happened in the past.

free will. Not compelled; self-directed; capable of making decisions without being compelled by outside influences.

frequency. The number of cycles of energy per second; frequency is calculated from wavelength by dividing the speed of light by the wavelength. $v = c\lambda^{-1}$ where v is the frequency, c is the speed of light and λ is the wavelength.

gemmules. Units of heredity shed by each cell in the discredited theory of pangenesis.

geometry, Lobachevskian. A geometry that is based on negatively curved planes, ie, saddle shaped. Parallel lines diverge infinitely.

geometry, Newtonian. A geometry based on flat planes. Parallel lines never meet and remain equidistant indefinitely.

geometry, Riemannian. A geometry based on positively curved planes, ie, spherical. Parallel lines meet in at least two points (because they are actually arcs on a sphere).

GIGO (Garbage In, Garbage Out). A problem with computer programs or models whereby incomplete or erroneous data input produces erroneous output.

god of the gaps. A derogatory term for a belief that all of science is incorporated in Genesis between or before the creation narrative.

Goldschmidt break. Barrier between species.

Gould, Steven Jay. Evolutionist who postulated punctuated evolution where long periods of stability are followed by short periods of rapid evolution of new species.

gradualism. Belief in slow change over long periods of time by a series of small changes.

gravity. The attraction of one body to another proportional to their masses and inversely proportional to the square of the distance between them by a proportionality constant. $F = G(m_1 m_2/d^2)$ where F is the force and G is the universal gravitational constant ($G = 6.67 \times 10^{-11}$ nt-m^2/kg^2).

hadron. A subatomic particle such as a proton or neutron that is composed of smaller particles such as quarks.

hagiography. Idealizing or idolizing biography (as of saints) - from *Hagiographa*, the third part of the Jewish scriptures.

hagiology. Literature dealing with venerated persons (such as saints).

HDTKT (How do they know that?). A question asked about the logical source of a statement.

heuristic. Starting with a preconceived idea and proceeding through experimentation or observation to conclusions.

histone. Any of the proteins that DNA is wrapped around.

homology. Similar forms, especially of fossils, attributed to common ancestors; relevant to classification.

Hubble Constant. The ratio between the distance and the speed of recession of redshifted objects expressed as Km/sec/Mpc meaning the speed in Km/sec per Megaparsec distance. Current value is 77Km/sec/Mpc +/- 15 percent.

icon. An object of uncritical devotion.

iconic. Of, related to or having the character of an icon.

iconoclasm. The doctrine, practice or attitude of an iconoclast.

iconoclast. One who attacks established beliefs or institutions.

iconography. Pictures representing a subject such as saints and their deeds.

iconology Study of icons or artistic symbolism.

Idealism. The theory that ultimate reality lies in a realm transcending phenomena, as opposed to materialism.

induction. The act, process or result or an instance of reasoning from a part to the whole, from particulars to generals, or from the individual to the universal.

insignificant or complex cause. One thing is assumed to cause another but it is only one, perhaps minor, part of a group of causes.

instinct. Behavior that is inherent and inherited and that does not require conscious thought or learning.

intron. Sections of DNA between known genes for which science has yet to determine functions; sometimes called "junk DNA."

interdisciplinary. Crossing into more than one field of study.

invidious. Tending to cause discontent, animosity, or envy.

IPCC. United Nations Intergovernmental Panel on Climate Change, a political organization employing scientists but guided by political operatives.

irrelevant conclusion. Logical Fallacy: An argument for one conclusion really proves a different one.

junk DNA. Also called introns; sections of DNA between known genes for which science has yet to determine functions.

kitch. Vulgar or imitation art; anything that is a cheap imitation of reality.

lepton. A subatomic particle that is indivisible such as electrons, mesons, bosons.

light year. The distance that light can travel in a vacuum in one year. The speed of light is 3×10^8 meters per second, so a light year is approximately 9.46×10^{12} kilometers or about 5.88×10^{12} miles, (9.46 trillion kilometers or 5.88 trillion miles).

Logarithm. Is the exponent of a number, usually 10, written 10^2. An exponent is the number of times minus one (n-1) that a number is multiplied by itself, or the fractional equivalent thereof.

logic. (Gk logikos = of reason; logos = reason)

1. a science that deals with the canons and criteria of validity of inference and demonstration: the science of the normative formal principles of reasoning.
2. a branch of semiotic especially syntactics.

logical. 1a: Relating to, in accordance with, or skilled in logic. 1b: Formally true or valid: analytic, deductive. 2: That which is in accordance with inferences reasonably drawn from events and circumstances.

logical fallacy. An argument failing to satisfy the conditions of valid or correct inference.

logical positivism. (or logical empiricism). A twentieth-century philosophical movement that holds that characteristically all meaningful statements are either analytic or conclusively verifiable or at least confirmable by observation and experiment and that metaphysical theories are therefore strictly meaningless.

MACHO. Massive Compact Halo Objects. One of the possible forms of dark matter. See also.

magic. A process with hidden technology or expertise whereby one thing is assumed to have happened but another has actually happened.

magical thinking. A process of drawing conclusions that seem to defy logic. A type of causal fallacy where events are linked in a type of cause and effect relationship that is undetectable or missing.

materialism. The philosophical doctrine that matter is the only reality and that everything in the world, including thought, will, and feeling, can be explained only in terms of matter as opposed to idealism.

mechanism. A philosophy that all natural phenomena, particularly life, can ultimately be explained through physics and chemistry, as opposed to vitalism.

media. Includes any print, electronic, or broadcast media such as movies, television, internet, newspapers, magazines, or books presenting news, entertainment, documentary, opinion or similar information to the general public.

megaparsec. 1,000,000 Parsec.

metamorphosis. The process whereby a living being transforms itself from one form to another. Example: caterpillar to butterfly.

metaphysical.
1. Of or relating to metaphysics.
2. Of or relating to the transcendent or supersensible.
3. Highly abstract or abstruse.

metaphysics. From Greek meta meaning beyond and phycika meaning physical, Beyond the physical; that which is transcendent or supersensible; also the underlying principles of a particular subject. (Science, at a fundamental level, is based on metaphysics because it originates in thought, which has metaphysical qualities.)

miracle. An event that defies the laws of logic and probability through an unknown factor or process, eg, God, that makes it appear that these laws have been broken.

mode. In statistics, the portion of the curve (value) that has the most data points.

naturalism. The belief that the natural world, known and experienced scientifically, is all that exists and that there is no supernatural or spiritual creation, control or significance.

natural theology. A theology that is based on observation of natural processes and not on divine revelation.

neo-Darwinism. Evolutionary theory reformulated to include later knowledge such as DNA.

non sequitur. (Latin: "it does not follow") An inference that does not follow from the premise.

Occam's razor. A principle of science that says, given more than one choice, the simplest solution is usually the correct one.

ontogenetic. Based on morphological characters; of, relating to, or appearing in the course of ontogeny.

ontogeny. The development or course of development of an organism.

ontology. (Gk - ont = to be; logikos = of reason)
1. A branch of metaphysics relating to the nature and relation of being.
2. A particular theory about the nature of being or the kinds of existence.

oxymoron. A combination of contradictory or incongruent words, eg, cruel kindness.

panchronic species. Species that have remained relatively unchanged through many geologic eras.

pangenesis. A discredited theory from the nineteenth century that each cell in the body sheds gemmules that migrate to reproductive cells to make up inherited traits.

paradox. A statement that is seemingly contradictory or opposed to common sense and yet may betrue; an argument that apparently derives self-contradictory conclusions by valid deduction from acceptable premises.

paradigm. A picture or view of reality into which all other facts and beliefs must fit.

parallax. See stellar parallax and spectroscopic parallax

parsec. The distance to an object with a stellar parallax angle of 1 second of arc, which is 3.26 light years or 30.857×10^{12} or 30.857 trillion kilometers (19.178×10^{12} or 19.178 trillion miles) distance.

pataphysics. A speculative theory which is itself based on other speculative theories. (Pataphor in Wikipedia). "Lopez has also asserted that far-reaching concepts such as string theory constitute a kind of mathematical pataphor, insofar as these concepts correspond to the pataphysical notion of "supposition built on supposition." In other words, as string theory is speculation based on ideas that are *themselves speculative* (in this instance, the theories of general relativity and quantum mechanics), string theory is not in fact physics, but pataphysics." Note that Wikipedia has since removed any reference to science or mathematics in later versions of this article.

peer review. A process where a panel of other scientists in the field, chosen by a publisher, reviews and approves of scientific reports. Peer review does not necessarily mean an unbiased assessment of the work, only that the reviewers don't strongly disagree with it or the methods employed.

perihelion. The nearest point in a planet's orbit to the sun.

phylogenetic. Based on natural evolutionary relationships; acquired in the course of phylogenetic development.

phylogeny. The evolution of a genetically related group of organisms as distinguished from the development of the individual organism. See also ontogeny.

physicalism. 1. A view that people are purely physical objects with no nonphysical component. 2. Thesis that the descriptive terms of scientific language are reducible to terms that refer to spatiotemporal things or events or to their properties.

plate tectonics. A theory that the crust of the earth is composed of plates that move relative to each other carrying continents with them. See also continental drift.

positivism. The philosophy that the only authentic knowledge is that based on actual sense experience. It excludes metaphysical speculation. Also logical positivism. The positivist view is sometimes referred to as a "scientistic ideology."

post hoc. Because one thing follows another it is assumed to cause it.

postmodernism. A belief that there is no absolute truth, that everything is relative and subjective; knowledge is an invention of the mind and may be different for different types of thinkers.

pragmatics. (Gk. Pragmatikos – skilled in law or business) A branch of semiotic that deals with the relation between signs or linguistic expressions and their users.

pragmatism. A practical approach to problems and affairs; the function of thought is to guide action.

premise. A proposition supposed or proved as a basis of an argument or inference.

probability. The likelihood or odds of an event occurring. An area of mathematics in which this is calculated.

propaganda. 1. The spreading of ideas, information or rumor for the purpose of helping or injuring an institution, a cause or a person. 2. Ideas, facts or allegations spread deliberately to further one's cause or to damage an opposing cause; also a public action having such an effect.

proxy data. Where one thing is measured in lieu of another, more difficult or impossible measurement and interpreted as meaning the other.

punctuationism. The belief in long periods of stability broken by brief periods of rapid evolution.

quark. A subatomic particle which is combined to form Protons and Neutrons, both classified as hadrons.

quintessence. A hypothetical energy field filling the universe that is changeable depending on the mass of objects within it.

rationalism. A theory that reason is in itself a source of knowledge superior to and independent of sense perception.

red herring. Logical Fallacy: Inserting another unrelated factor intended to throw off the opposition rather than address the issue.

redshift. The shift to longer (redder) wavelengths in the spectrum of light from stars and galaxies that is thought to be caused by the apparent speed at which they are moving away from Earth. To be more precise, it is measured by the shift in spectral gaps (dark lines) in the spectrum caused by absorption of light at given wavelengths by certain elements.

reductionsim. A procedure or theory that reduces complex data or phenomena to simple terms.

ribosome. Organelle inside a cell that uses directions from RNA to facilitate construction of proteins from amino acids.

relativism. A theory that knowledge is relative to the limited nature of the mind and the conditions of knowing; a view that ethical truths depend upon the individuals and groups holding them.

ridicule. Malicious belittling; contemptuous scorn, often ironically expressed.

saltationism. Sudden appearance of new species sometimes called "hopeful monster."

science. The pursuit of truth about the predictable, repeatable and measurable aspects of the universe with which we can or could conceivably interact.

Science. A weekly publication of AAAS, American Association for the Advancement of Science, containing both peer reviewed scientific papers and popular science news articles.

science of the gaps. That which occurs when science fills in a lack of knowledge with clever stories.

semantics. Branch of semiotics dealing with the relations between signs and what they refer to and including theories of denotation, extension, naming and truth;

especially connotative meaning; the exploitation of connotation and ambiguity, as in propaganda.

semiotic. (Gk - semeiotikos - observant of signs) A general philosophical theory of signs and symbols that deals especially with their function in both artificially constructed and natural language and comprises syntactics, semantics, and pragmatics. Semiotical, Semiotician.

shibboleth. Catch phrase or slogan.

sine qua non. (Latin: without which not) An absolutely indispensable or essential thing.

skew. Statistical term meaning the Bell curve of a distribution is asymmetrical or distorted.

specious. An argument that has a false appearance of truth or genuineness. Speciousness, speciosity, speciously.

spectroscopic parallax. Not true parallax; really fitting stars into a brightness and color vs. size relationship; not very precise with many exceptions.

SQR (Status quo regurgitators). A term for sophisticate level of belief and understanding of science. Those who espouse the standard theories of science without question.

standard candle. A stellar object of assumed known brightness or other characteristics used to gauge interstellar and intergalactic distances. There is a series of standard candles to cover the vast distances between galaxies. Counting from nearest to farthest they are: radar, stellar parallax, spectroscopic "parallax" (really fitting stars into a brightness and color vs. size relationship), Cepheid variables, Tully-Fisher galaxy brightness relationship, Type 1a supernova, and redshift alone.

statistics. A branch of mathematics dealing with the collection, analysis, interpretation and presentation of masses of numerical data.

status quo. The way things are; the existing state of affairs.

stellar parallax. The angular distance a star appears to move against the background due to the motion of the Earth in its orbit.

straw man. Logical Fallacy: Arguing against a weaker proposition to imply winning a stronger one.

subjectivism. A theory that limits knowledge to conscious states and elements; a doctrine that the supreme good is a subjective experience.

syllogism. Deductive reasoning; a deductive scheme of a formal argument consisting of a major and a minor premise and a conclusion.

synergy. Combined effect of two or more factors. Positive synergy is where the total effect of two or more factors results in a larger effect than the sum of the effects of each factor alone. Negative synergy is where the total effect is less than the sum of each factor – as if they partially cancel each other out.

syntactics. (Lat. Syntacticus, Gk syntaktikos: arranging together) A branch of semiotic that deals with the formal relations between signs or expressions in abstraction from their signification and their interpreters. Related: Syntactic – Of or relating to, or according to the rules of syntax or syntactics.

syntax.
1. Connected or orderly system of arrangement.
2. The way in which words are put together to form phrases, clauses or sentences.

synthetic.
1. Attributing to a subject a predicate that is not part of the meaning of that subject
2. Empirical
3. Not resulting in a contradiction upon being negated.

tautology. True by virtue of its logical form alone; a needless repetition of an idea, statement or word.

teleological. Exhibiting or relating to design or purpose esp. in nature <~ argument for God's existence>.

teleologist. A specialist or believer in teleology.

teleology. (Gk tele or telos = end, purpose.)
1. The study of evidences of design in nature; a doctrine that ends are immanent in nature;
2. A doctrine explaining phenomena by final causes;
3. The fact or character attributed to nature or natural processes of being directed toward an end or shaped by a purpose;
4. Use of design or purpose as an explanation of natural phenomena.

tetchy. Irratibly or peevishly sensitive; touchy.

theomachy. Opposition to god(s) or divine will; battle or strife among the gods.

theomonism. Belief that one divine spirit rules the universe.

theomorphic. Formed in the image of the deity.

theonomous. Subject to God's authority.

theopantism. Mystical doctrine that God is the sole reality.

theophany. Something manifesting or revealing deity.

theophilanthopist. Member of a deistic society established in Paris during the period of the Directory aiming to institute in place of Christianity, which had been officially abolished, a new religion affirming a belief in the existence of God, in the immortality of the soul and in virtue.

theory. A plausible or scientifically acceptable statement offered to explain observed physical phenomena.

transcendent. That which lies outside or beyond the universe or material world with which we can interact.

transcendental. Of or related to a super sensible quality; a belief in a transcendent quality to the universe.

Transcendentalism. A nineteenth-century idealistic movement that claimed personal inspiration as the only way to God, which was espoused by such noted figures as Ralph WaldoEmerson and Henry David Thoreau.

true.
1. In accordance with the actual state of affairs.
2. Conformable to an essential reality.
3. Ideal, essential.
4. Being that which is the case rather than what is manifest or assumed.
5. Consistent.

truism. A self-evident truth, especially one too obvious or unimportant to mention.

truth.
1. The state of being the case.

2. The body of real things, events, and facts: actuality.
3. The body of true statements or propositions. SYN: veracity, verity, verisimilitude

utilitarianism. A philosophy in which virtue is that which produces the most good for the most people.

utopianism. (and scientific utopianism) Belief in the perfectibility of human societies (socialist).

vacuum. Space devoid of gases or other content. Partial vacuum contains tiny amounts of matter.

vacuum energy. Theoretical energy that arises from empty space and caused by the annihilation of spontaneously produced viral pairs of particles and antiparticles.

verifiable. Capable of being shown to be true or probably true. One of the conditions defining scientific theories. See also falsifiable.

vitalism. (Lat. Vitalis = of life; vita = life)
1. A doctrine that the functions of living organism are due to a vital principle distinct from the physico-chemical forces.
2. A doctrine that the processes of life are not explicable by the laws of physics and chemistry alone and that life is in some part self-determining.

vitalist. A believer in vitalism.

wavelength. The length of a wave of energy in centimeters. Wavelength is related to frequency by dividing the speed of light by the wavelength. $v = c\lambda^{-1}$ where v is the frequency, c is the speed of light and λ is the wavelength.

Wavenumber. The reciprocal of the wavelength (λ^{-1}).

weasel words. Qualifiers included in a statement to allow for the possibility that the statement may not be entirely true or proven. Examples: *probably, is believed to be, possibly, presumably, we think, sometimes.*

WIMP. Weakly Interacting Massive Particle. One of the possible forms of dark matter.

wrong direction. An argument in which the cause and effect are reversed.

SELECTED READINGS

EVOLUTION

- Behe, Michael J. *Darwin's Black Box.* The Free Press, 1996.

- Darwin, Charles. *On the Origin of Species by Means of Natural Selection or The Preservation of Favoured Races in the Struggle for Life,* first edition, London, John Murray, 1859; Dover Thrift Editions, Mineola, NY, 2006, an unabridged republication on the original work; sixth edition, 1871 Kindle edition: Published by B&R Samizdat Express.

- Dembski, William A. editor. *Darwin's Nemesis, Phillip Johnson and the Intelligent Design Movement.* Downers Grove, IL: InterVarsity Press, 2006.

- Dembski, William A. *The Design Inference, Eliminating Chance through Small Probabilities.* Cambridge University Press, 1998.

- Gould, Stephen Jay. *The Flamingo's Smile.* New York: W. W. Norton and Company, 1985.

- Gould, Stephen Jay. *Wonderful Life,* New York: W. W. Norton and Company, 1989.

- Johnson, Phillip E. *Darwin on Trial.* Washington, DC: Regnery Publishing, Inc., 1991.

- Malthus, Thomas. *An Essay on the Principles of Population.* London: J. Johnson, 1798.

- Mivart, St. George F.R.S. *On the Genesis of Species.* London: Macmillan And Co., 1871. Kindle Edition, 2007.

- Sagan, Carl. *Dragons of Eden.* New York: Ballantine Books, 1977.

- Wells, Jonathan. *Icons of Evolution, Science or Myth?* Washington, DC: Regnery Publishing, Inc., 2000.

- Wells, Jonathan. *The Politically Incorrect Guide to Darwinism and Intelligent Design* Washington, DC: Regnery Publishing, Inc., 2006.

LIFE

- Alberts, Bruce, Dennis Bray, Karen Hopkin, Alexander Johnson, Julian Lewis, Martin Raff, Keith Roberts, Peter Walter. *Essential Cell Biology*, second edition. Garland Science, 2004.

- Campbell, Neil A. *Biology*, fourth edition. The Benjamin/Cummings Publishing Company, Inc., 1996.

- Crick, Francis. *Life Itself, Its Origin and Nature*, New York: Simon and Schuster, 1981.

- Davies, Paul. *Are We Alone?* New York: Basic Books, 1995.

- Davies, Paul. *The Fifth Miracle*, New York: Simon and Schuster, 1999.

- de Duve, Christian. *Vital Dust*, New York: Basic Books: Perseus Books, LLC, 1994.

- Lehninger, Albert L. *Principles of Biochemistry*, first edition. Worth Publishing, Inc., 1982.

- Lehninger, Albert L., David L. Nelson, Michael M. Cox, *Principles of Biochemistry*, second edition. Worth Publishing, Inc., 1993.

- Lewis, Rikki, Douglas Gaffin, Marielle Hoefnagels, Bruce Parker, *Life*, fifth edition, New York: McGraw Hill, 2004.

- Shapiro, Robert. *Origins*, New York: Bantam Books, 1986.

- Shapiro, Robert. *Planetary Dreams.* Hoboken, NJ: John Wiley and Sons, Inc., 1999.

- Stryer, Lubert. *Biochemistry*, second edition, 1981.

- Ward, Peter D., Donald Brownlee. *Rare Earth, Why Complex Life is Uncommon in the Universe*, Gottingen, Germany: Copernicus Books, 2000.

COSMOLOGY AND ASTRONOMY

- Ashmore, Lyndon. *Big Bang Blasted*, North Charleston, SC: BookSurge LLC, 2006 (alternative cosmology).

- Chaisson, Eric, Steve McMillan, *Astronomy Today*, fifth edition, Upper Saddle River, NJ: Pearson Prentice Hall, 2005.

- Disney, Michael. *The Hidden Universe*, New York: MacMillan, 1984.

- Eddington, Arthur. *The Expanding Universe*, Cambridge University Press, 1933.

- Hawking, Stephen W. *A Brief History of Time*. Dove Books on Tape, Inc., 1988

- Hoyle, Fred. *Astronomy*. Wingdale, NY: Crescent Books, Inc., 1962.

- Jeans, James. *The Universe Around Us*. New York: Macmillan Company, 1929.

- Lerner, Eric. *The Big Bang Never Happened*. New York: Times Books: Random House, 1991 (alternative cosmology).

- Radcliffe, Hilton. *The Static Universe*, Montreal, QB, C. Roy Keys, Inc., 2010 (alternative cosmology).

- Rosen, Joe. *The Capricious Cosmos*, New York: Macmillan, 1991.

- Shu, Frank. *The Physical Universe*. Herndon, VA: University Science Books, 1982.

- Taff, Laurence G. *Celestial Mechanics*. Hoboken, NJ: Wiley-Interscience, 1985.

- Witt, Terence. *Our Undiscovered Universe*. Indialantic, FL: Aridian Publishing Corp., 2007. (alternative cosmology)

- Young, Charles A. *The Elements of Astronomy, A Text Book*, revised edition, Boston: Ginn & Company, 1889.

- Young, Charles A. *General Astronomy*, Boston: Ginn & Company, Boston, 1897.

PARTICLE PHYSICS & QUANTUM MECHANICS

- Werner Heisenberg, *Physics and Philosophy, The Revolution in Modern Science*, Introduction by Paul Davies, Harper & Rowe, New York, 1962

- Albert Einstein, *The Theory of Relativity & Other Essays*, Philosophical Library, Inc., 1950

- Richard F. Feynman, *The Character of Physical Law*, The M. I. T. Press, 1990

- Richard F. Feynman, *QED, The Strange Theory of Light and Matter*, Princeton University Press, 1985

CLIMATE AND ENVIRONMENT

- Crichton, Michael. *State of Fear*. New York Avon Books: Harper Collins Publishers, 2004 (a novel using real climate data).

- Horner, Christopher C. *The Politically Incorrect Guide to Global Warming and Environmentalism*. Washington, DC: Regnery Publishing, Inc., 2007.

- Horner, Christopher C. *Red Hot Lies*, Washington, DC: Regnery Publishing, Inc., 2008.

- Inhofe, Sen. James. *The Greatest Hoax*, Washington, DC: WND Books, 2012.

- Lomborg, Bjorn. *Cool It: The Skeptical Environmentalist's Guide to Global Warming*. New York: Alfred A. Knopf, 2008.

- Lomborg, Bjorn. *The Skeptical Environmentalist*, Cambridge, UK: Cambridge University Press, 2001.

- Plass, Gilbert. *The Carbon Dioxide Theory of Climatic Change*. Baltimore, MD: Johns Hopkins University, 1956.

- Spencer, Roy W. *The Great Global Warming Blunder*, New York, NY: Encounter Books, 2010.

- Sussman, Brian. *ClimateGate*, Washington, DC: WND Books, 2010.

- Tennesen, Michael. *The Complete Idiot's Guide to Global Warming*. New York: Alpha: Penguin Group Inc., 2004.

- Tyndall, John. *Contributions to Molecular Physics in the Domain of Radiant Heat*. Cambridge, UK: Cambridge University Press, 1872.

- Wayne, Richard P. *Chemistry of Atmospheres*, second edition. Clarendon Press: Oxford Science Publications, 1991.

SCIENCE AND PHILOSOPHY

- Baker, Henry. *Of Microscopes and the Discoveries Made Thereby*, London, 1785

- Barker, Stephen F. *The Elements of Logic*, fifth edition, New York: McGraw-Hill, 1989.

- Beveridge, W. I. B. *The Art of Scientific Investigation*, New York: W. W. Norton & Company, Inc., 1950.

- Copernicus, Nicolaus. *De Revolutionibus Orbium Coelastium, (Of the Revolutions of the Celestial Spheres)*, 1543.

- Dembski, William A. *The Design Inference, Eliminating Chance Through Small Probabilities*, Cambridge Studies in Probability, Induction, and Decision Theory, 1998, 2007.

- Einstein, Albert. *Albert Einstein: Philosopher-Scientist*, translated from the German by Paul Arthur Schlipp, Vol. 1 of The Library of Living Philosophers, first printing 1949.

- Faraday, Michael. *Experimental Researches in Electricity*, three volumes. London: J. M. Dent & Sons, Ltd. 1922, PDF copy, 2013, from www.forgottonbooks.com

- Galilei, Galileo. *Sidereus Nuncius (The Starry Messenger)*, 1610.

- Galilei, Galileo. *The Assayer*, 1623.

- Galilei, Galileo. *Dialogue on the Two Chief Systems of the World*, 1632.

- Huff, Darrell. *How to Lie with Statistics*, New York: W. W. Norton & Company, 1954, now its 48th printing.

- Jones, Gerald Everett. *How to Lie with Charts*, Lincoln, NE: Author's Choice Press, 1995, 2000.

- Kuhn, Thomas S. *The Structure of Scientific Revolutions*, 2nd ed. Enlarged. Chicago, IL: University of Chicago Press, 1970.

- Monmonier, Mark. *How to Lie with Maps*, Second Edition, Chicago, IL: University of Chicago Press, 1996.

- Nute, Donald. *Methods for Critical Thinking*, third edition, Old Tappan, NJ: Pearson Custom Publishing, 2005 with material taken from Irving M. Copi, Carl Cohen, *Introduction to Logic*, tenth & eleventh editions, 1998, 2002.

- Popper, Karl. *The Logic of Scientific Discovery*. New York NY: Basic Books. 1959.

GENERAL AND MISCELLANEOUS INFORMATION

- Bethell, Tom. *The Politically Incorrect Guide to Science*, Washington, DC: Regnery Publishing, Inc., 2005.

- Box, George E. P., William G. Hunter, J. Stuart Hunter, *Statistics for Experimenters*. New York: John Wiley & Sons, Inc., 1978

- Bryson, Bill. *A Short History of Nearly Everything*, New York: Broadway Books, 2003.

- Gross, Ronald. *Independent Scholar's Handbook*, Boston: Addison-Wesley Publishing Company, 1983.

- Hogben, Lancelot. *Mathematics for the Million*, New York: W. W. Norton & Company, 1968, (quick reference).

- Lyell, Charles. *Principles of Geology*, London. John Murray, 1830.

- Maddox, John. *What Remains to be Discovered*. New York: The Free Press/ Simon & Schuster, 1998.

- Marx, Karl. *Das Kapital*, Otto Meissner, 1867.

- Primack, Alice Lefler. *Finding Answers in Science and Technology*, New York: Van Nostrand Reinhold, 1984.

- Spiegel, Murray R. *Probability and Statistics*, Schaum's Outline Series in Mathematics. New York: McGraw-Hill, 1975 (quick reference).

HISTORY

- Barton, David. *The Jefferson Lies*, Nashville, TN: Thomas Nelson, 2012.

- Irving, Washington. *A History of the Life and Voyages of Christopher Columbus.* New York: G.P. Putnam's Sons,1828.

- Morison, Samuel Eliot. *Admiral of the Ocean Sea, A Life of Christopher Columbus*, Boston: Little, Brown and Company, 1942.

FICTION AND HUMOR

- Adams, Douglas. *Hitchhiker's Guide to the Galaxy*. New York: Random House, 1979.

- Twain, Mark. *Life on the Mississippi*. New York: Harper and Brothers, 1917.

- Ward, Artemus. *The Complete Works Of Artemus Ward (Charles Farrar Browne), Part I. Essays, Sketches, And Letters*. New York: G. W. Dillingham Co., 1898.

REFERENCE BOOKS

- *Encyclopedia Britannica*, fifteenth edition, 1983; and online www. britannica.com

- Wikipedia www.wikipedia.com Note: because this source is written by users and evolves, the information may be of questionable validity and may not be unbiased. For that reason, I have cross-checked information from this source with other sources or gone to the cited original sources.

- *Catholic Encyclopedia* at http://www.newadvent.org/

- *McGraw-Hill Concise Encyclopedia of Science and Technology.* New York: McGraw-Hill, 1984.

- *Van Nostrand's Scientific Encyclopedia*, fifth edition, New York: Van Nostrand Reinhold, 1968.

ENDNOTES

INTRODUCTION

1. Paradigm – an overall picture or view or reality into which all other facts and beliefs must fit.

2. Teleological – exhibiting or relating to design or purpose esp. in nature <~ argument for God's existence>

3. *The Capricious Cosmos* by Joe Rosen, 1991, Macmillan, New York.

4. Popularized in *A History of the Life and Voyages of Christopher Columbus*, 1828, by Washington Irving.

5. Bible: Isaiah 11:12 – "And he shall set up an ensign for the nations, and shall assemble the outcasts of Israel, and gather together the dispersed of Judah from the four corners of the earth." (King James version)

6. Bible: Revelation 7:1 – "And after these things I saw four angels standing on the four corners of the earth, holding the four winds of the earth that the wind should not blow on the earth, nor on the sea, nor on any tree." (King James version)

7. Bible: Isaiah 40:22 – "It is he that sitteth upon the circle of the earth, and the inhabitants therof are as grasshoppers; that streatcheth out the heavens as a curtain, and spreadeth them out as a tent to dwell in." also Job 26: 7 – "He stretcheth out the north over the empty place, and hangeth the earth upon nothing." (King James version)

8. Oxymoron – a combination of contradictory or incongruous words such a "cruel kindness"

9. A takeoff on the God of the Gaps claim of atheists to characterize any belief in a creator or design.

10. Except for the original verbal presentation of a paper on the subject to the Linnean Society.

CHAPTER 1

1. Paradigm – A picture or view of reality into which all facts and beliefs must fit.

2. Dogma –established opinion put forth as authoritative, especially without adequate grounds.

3. Oxymoron – a combination of contradictory or incongruent words, eg, cruel kindness.

4. Devil's advocate – one who champions a contrary view for the sake of argument.

5. Logical fallacy – an argument failing to satisfy the conditions of valid or correct inference.

6. Metaphysics – (from Greek meta meaning beyond and phycika meaning physical) beyond the physical; that which is transcendent or supersensible; can also mean the underlying principles of a particular subject. Science, at a fundamental level, is based on metaphysics because it originates in thought, which has metaphysical qualities.

7. Forensics – piecing together evidence to support a hypothesis about what probably happened in the past.

8. Heuristic – starting with a preconceived idea and proceeding through experimentation or observation to conclusions.

9. Anecdote – a short narrative of an interesting, amusing, or biographical incident whose reliability is not certain.

10. *How to Lie with Statistics* by Darrell Huff, New York: Penguin Books 1973; Reissue edition, New York: W. W. Norton & Company, 1993. First published by W. W. Norton & Company, 1954, now in its 48th printing.

11. "Just think: The challenges of the disengaged mind," *Science,* 4 July 2014: Vol. 345 no. 6192 pp. 75-77 DOI:10.1126/science.1250830, Timothy D. Wilson, David A. Reinhard, Erin C. Westgate, Daniel T. Gilbert, Nicole Ellerbeck, Cheryl Hahn, Casey L. Brown, Adi Shaked.

12. Combined effect of two or more factors. Positive synergy is where the total effect of two or more factors results in a larger effect than the sum of the effects of each factor alone. Negative synergy is where the total effect is less than the sum of each factor – as if they partially cancel each other out.

CHAPTER 2

1. Status quo – the way things are; the existing state of affairs.

2. Transcendent – that which lies outside or beyond the universe or material world with which we can interact. Some believe that life itself, thought and existence are transcendent qualities. Not to be confused with Transcendentalism, a nineteenth-century idealistic movement that claimed personal inspiration as the only way to God, which was espoused by such noted figures as Ralph Waldo Emerson and Henry David Thoreau.

3. Ad hoc – for this particular purpose; special; created on the spur of the moment, impromptu.

4. Douglas Adams, *Hitchhiker's Guide to the Galaxy*, 1979.

5. Michael Disney, *The Hidden Universe*, 1984.

6. Evangelical – zealously enthusiastic about promoting the spread of the gospel, (in this case, the gospel of status quo scientific dogma).

7. Materialism – a theory that matter is the only reality and that everything in the world, including thought, will, and feeling, can be explained only in terms of matter, (as opposed to idealism).

8. Naturalism – the belief that the natural world, known and experienced scientifically, is all that exists and that there is no supernatural or spiritual creation, control or significance. Natural theology is belief that is based on observation of natural processes as opposed to divine revelation.

9. Peer reviewed does not mean the content is flawless or scientifically rigorous, only that the reviewers, chosen by the publisher, don't strongly disagree with the stated methods or conclusions of the author.

10. *National Geographic*, January 2007, "Flight of Fancy" by Michael Klesius, photographed by Luis A. Mazariegos.

11. See http://www.eastangliaemails.com/ for searchable database of the emails. "Hacked E-Mail Is New Fodder for Climate Dispute" by Andrew Revkin, New York Times, November 21, 2009.

12. IPCC is the United Nations Intergovernmental Panel on Climate Change.

13. See http://foia2011.org for searchable database of the earlier and latest emails. *The Telegraph*, James Delingpole blog. *The Blaze*, Liz

Klimas, "Environment Climategate Part II? New Leaked Emails Again Point to Global Warming Activism."

CHAPTER 3

1. Karl Popper, *The Logic of Scientific Discovery*, Basic Books, New York NY, 1959.

2. Panchronic - throughout time or across time.

3. ad hoc - for this particular purpose; special; created on the spur of the moment, impromptu.

4. Wikipedia article on Crabtree's Bludgeon: "…appear in the Crabtree Orations, a set of satirical academic commentaries attributed to the fictitious poet, Joseph Crabtree. Crabtree was originally a creation of Sir James Sutherland, and was later maintained by the Crabtree Foundation."

5. *Science* is a weekly publication of AAAS, American Association for the Advancement of Science, containing both peer reviewed scientific papers and popular science news articles.

6. In searching the *Science* archives on line, I have failed to locate this article, perhaps because it has been removed after the retraction. All I could find was a news article skeptically describing such studies.

7. In reality, this is a simplification of the theory that postulated that both the sun and the earth revolved around a center of gravity. Although this is true, the vastly greater mass of the sun makes it almost moot. The difference between an unmovable sun and a sun- Earth system where each gravitationally influences the other is so small as to be almost as negligible as the pull on the Earth by a man's mass.

8. According to the *Catholic Encyclopedia*, "Bruno was not condemned for his defense of the Copernican system of astronomy, nor for his doctrine of the plurality of inhabited worlds, but for his theological errors, among which were the following: that Christ was not God but merely an unusually skillful magician, that the Holy Ghost is the soul of the world, that the Devil will be saved, et cetera,-" For more details see Catholic Encyclopedia at http://www.newadvent. org/cathen/06342b.htm

9. The Inquisition was the arbiter of all printed matter in Catholic-dominated Europe. Before any work could be distributed, it had

to pass muster with the Inquisition. Often, they asked authors to insert or change wording to align it with Catholic doctrine. The Inquisition of the Catholic Church should not be confused with the later, extremely cruel Spanish Inquisition, which was run by local Spanish clerics.

10. *The Elements of Astronomy, A Text Book*, p. 347, by Charles A. Young, 1889, 1897. See also *General Astronomy* by the same author.

CHAPTER 4

1. Media here means any print, electronic or broadcast media such as movies, television, internet, newspapers, magazines, or books presenting news, entertainment, documentary, opinion or similar information to the general public.

2. "Weasel words" as used here follows the original definition that is a way to say something imprecisely so that it appears precise but with wiggle room. An example is "up to 50 percent better," which can mean any amount up to 50 percent, when it appears to promise 50 percent improvement. Note that in some circles weasel words have come to be defined as an emotionally charged label used to poison the well as, for instance, with guilt by association. Please, do not confuse the two usages.

3. A priori – from the beginning.

4. Infinite improbability drive is from *Hitchhiker's Guide to the Galaxy*, a satirical novel by Douglas Adams.

5. COWDUNG – the COnventional Wisdom of the DomiNant Group, Michael Disney, *The Hidden Universe*.

6. Macroevolution – that which gives rise to whole new species, as opposed to micro-evolution that describes variations within a single species. Microevolution is well known and has been applied successfully by breeders of everything from petunias to dogs.

7. Specious – having a false appearance of truth or genuineness.

8. Non sequitur – (Latin: it does not follow) an inference that does not follow from the premise.

9. Skew is a statistical term meaning the Bell Curve of a distribution is asymmetrical or distorted.

10. Mode is the portion of the Bell Curve that has the most data points, so a bimodal curve has two peaks.

11. *How to Lie with Statistics*, Darrell Huff (W. W. Norton & Company). *How to Lie with Charts*, Gerald Everett Jones (iUniverse). *How to Lie with Maps*, Mark Monmonier (University Of Chicago Press).

12. Left-handed amino acids – amino acids have two possible spatial orientations, called right- and left-handed depending on the way they bend light. Synthetic amino acids contain a mixture of both right- and left-handed amino acids, but biological systems contain only the left-handed versions. Biological systems also use only right-handed sugars. Only 20 amino acids are used by biological systems, although many other synthetic amino acids are possible. How these things came to be and the reasons for them are unclear.

13. *The Design Inference: Eliminating Chance through Small Probabilities*, William A. Dembski (Cambridge University Press, 1998).

PART 2

1. From *Hitchhiker's Guide to the Galaxy* by Douglas Adams.

CHAPTER 6

1. Standard model used here should not be confused with the Standard Model of particle physics, although the two may be interconnected.

2. "A Galaxy when Galaxies Were Young" by Jessica Kloss, *Sky and Telescope*, January 27, 2011.

3. Million million = trillion = 1 followed by 12 zeros.

4. Million million billion = a 1 followed by 21 zeros.

5. This is a simplification; there is a theory that this rate is accelerating.

6. Isaac Asimov, "... Star X, spectral class G0, 4 **planets** plus **debris...**" from "By Jove!" in *View from a Height* (1963).

7. HDTKT – How Do They Know That

8. Douglas Adams, *Hitchhiker's Guide to the Galaxy*.

9. Note that for any number x $10^{\#}$ a positive superscript means to move the decimal point that number of places in the original number to the right – a very large number. A negative superscript means to move the decimal point that number of places to the left – a very small number. Example: $2 \times 10^6 = 2{,}000{,}000$ and $2 \times 10^{-6} = 0.000002$

10. Pluto was recently downgraded from a planet to a minor planet.

11. Extremophile is a term applied to microbes on Earth that live in extreme conditions such as very hot or cold, very salty, acidic or alkaline, low or no oxygen, or where they get their energy from mineral sources instead of the sun.

12. Without greenhouse gases the Earth would be frozen solid and life would probably consist of a few extremophile microbes.

13. From record extremes of -89.2°C (-128.6°F) at Vostok Station, Antarctica to 56.7°C (134°F) at Death Valley, California.

14. Soviet Union Kola super deep borehole project 1970 to 1989.

15. Mariana Trench, Challenger Deep is 10.9km (6.8 mi.)

16. Water can remain liquid at higher temperatures due to high pressure at these depths.

17. Since classification has divided "bacteria" into two types, bacteria and archaea, for simplicity I will use the word microbe to mean either or both.

18. PSI means pounds per square inch.

CHAPTER 7

1. Note that 10^{27} is a 1 with 27 zeros after it and 10^{25} is a 1 with 25 zeros after it, and that 10^{27} is 100 times 10^{25}.

2. This is a simplification, but in essence is true. It is based on Avogadro's number, 6.0221×10^{23}, which is the number of atoms contained in one gram-molecular weight or mole of an element.

3. Carbon-14 also occurs in fossil fuels and crystalline minerals such as diamond. The source is uncertain, but radioactive elements may create it continually, such as when radioactive radium-223 decays to form carbon-14 and lead-209.

4. Approximately 10.5 half-lives. Each half-life halves the amount of carbon-14 remaining, so that would result in approximately 9.7×10^{-4} of the original amount of carbon-14 remaining.

5. Helium is smaller in size, but approximately four times the mass of hydrogen. Its smaller size is due to the electrons being more tightly held by the two protons in the nucleus than those of hydrogen which has only one proton. Additionally, hydrogen is usually found as a molecule composed of two hydrogen atoms (H_2) or combined with other elements.

6. Simplified version, since hydroxyl ions (OH⁻) are also produced.

7. Sublime means the solid transitions directly into a gas without first becoming a liquid.

8. The familiar hydrochloric acid, (aka muriatic acid), liquid is actually a gas dissolved in water.

9. Base pair refers to the paired molecules AT or GC as rungs of a twisted ladder shaped molecule with sugar phosphates as the sides. A = adenine; T=thymine; G=guanine; C=cytosine.

10. Chromosome is a large section of DNA carrying many genes.

11. Amino acids occur in two forms, the D and L isomers. D means dextro or right-handed and L means Levo or left-handed, referring to the way they bend polarized light. Only the L isomers of the 20 amino acids are used by living things. Similarly, sugars occur as D and L, but only the D form is used by living things.

12. Ribosomes are organelles that build proteins from amino acids following instructions from mRNA.

13. Mitochondria are organelles that produce energy for the cell. Each cell contains many mitochondria which have their own DNA and replicate independently of cell division.

14. Plastids are small membrane enclosed structures with specific functions such a chloroplasts which use energy from the sun, water and carbon dioxide to build sugars.

15. "Ever-Bigger Viruses Shake the Tree of Life," *Science*, Vol. 341, 19 July 2013, p. 226.

16. Ubiquitous – present everywhere.

17. Archaea is a recent change in nomenclature separating bacteria into two domains. Many archaea are extremophiles with internal structures or systems that are different from bacteria.

18. Symbiotic is a relationship between organisms that benefits both organisms, which are called symbionts.

19. Ediacaran formation in Western Australia.

20. Endo means internal so an endosymbiont is one that lives inside its host.

21. Cyanobacteria can exist in either aerobic or anaerobic conditions. Under anaerobic conditions, only the most primitive part of photosynthetic mechanism is used.

CHAPTER 8

1. Example: Waldensians were a group professing poverty, preaching, and opposed to practices such as image worship, relics, pilgrimages and intercession of the saints, that were persecuted by the Roman Catholic Church beginning in the twelfth century.

2. Originally from the Italian word *umanista* for a teacher of classical Greek and Latin beginning in the fourteenth century and promoted by Petrarch.

3. Note that the word *materialism* has been corrupted by the left to mean living for gain of material things, not a philosophy that denies any spiritual aspects. Reinterpretation of words is a common practice of the left that blurs real meanings.

4. *An Essay on the Principles of Population*, 1798, predicted starvation because populations were increasing exponentially while food supplies were increasing arithmetically. This philosophy assumed no improvements above subsistence-level farming, no development of more prolific and disease-resistant crops or mechanical means to increase production, and no other factors that would limit population such as disease and war.

5. Dialectical materialism supposedly progresses from thesis (original idea) through antithesis (opposition) to synthesis (final form).

6. Bourgeois (originally a resident of a town or burgh) is defined in Marxism as the (supposed oppressor) upper and middle classes as opposed to the proletariat defined as the (supposedly oppressed) lower classes. Proletariat is originally from the Latin *proletarius,* for citizens lacking property that were exempted from taxes and military service and could only contribute to the state by having children. This assumes that there is a strict class order rather than a fluid classless society whereby individuals assume ever changing positions based on effort and ability.

7. Capitalism was coined by early Socialists from capital, which originally meant head and later meant property or money.

8. Dictatorship of the proletariat is really a dictatorship by elites with special privileges over the masses, which are tightly controlled.

9. Only colony animals such as ants, bees, and a few rare vertebrates behave like that without coercion.

10. To be a Christian is to believe in Jesus Christ, repent of sins and rely on Him. Unlike all other religions, it is a religion of faith, not works. The Christian does good works not to ensure his salvation, but to emulate Jesus, follow His teachings and please God, all done out of gratitude for salvation already gained through simple faith.

11. A priori means presumed from the beginning; self-evident; intuitively obvious.

12. Eugenics is a "science" that deals with the improvement (as by control of human mating) of hereditary qualities of a race or breed. Systematized by Francis Galton, Charles Darwin's half-cousin, in the late nineteenth century, in which he advocated controlled breeding to prevent mankind from falling into mediocrity by regression towards the mean. This system was later used by the Nazis (National Socialist Party) in their pursuit of the master race, and was used to justify the elimination of Jews and other "undesirables."

13. Straw man argument is one where an easily defeated weaker premise is substituted for the real opposition view in order to appear to win the argument, ie, the author attacks an argument different from (and weaker than) the opposition's best argument

14. Ex nihilo is Latin for "from nothing"

15. Complete title and subtitles of the book is *On the Origin of Species by Means of Natural Selection or the Preservation of Favoured Races in the Struggle for Life*, published by John Murray, London, 1859.

16. Introduction, paragraph 8, describing Chapter 3, Struggle for Existence.

17. Dogma - established opinion put forth as authoritative, especially without adequate grounds.

18. Alfred Wegener, 1912, and earlier proponents.

19. Presented as "On the Tendency of Species to form Varieties; and On the Perpetuation of Varieties and Species by Natural Means of Selection." It was composed of two papers, Wallace's "On the Tendencies of Varieties to Depart Indefinitely from the Original Type" and Darwin's "Abstract Extract from an Unpublished Work on Species" along with "Abstract of a Letter to Asa Gray" (to establish primacy).

20. *On the Origin of Species*, first edition, Introduction, paragraph 3 & 4.

21. Example: *The Darwin Conspiracy: Origins of a Scientific Crime*, Roy Davies, 2008.

    Rebuttal: "There is no Darwin Conspiracy," Todd Charles Wood, Center for Origins Research and Education, Bryan College, Quote from the Introduction: "In his recent book, *The Darwin Conspiracy*, former BBC producer Roy Davies argued that Darwin's theory of evolution by natural selection was actually stolen from a variety of sources. According to Davies, Darwin began by appropriating natural selection from Edward Blyth and Patrick Matthew and concluded by stealing Alfred Russell Wallace's principle of divergence, all the while attempting to conceal his intellectual theft. Davies speculated that the guilt from these academic crimes was the source of Darwin's chronic illness." *Answers Research Journal* 2 (2009): 11–20.

CHAPTER 9

1. Recent unconfirmed work on certain genetic markers suggests that the whale is more closely related to the hippopotamus than other land animals from which it is assumed to have arisen.

2. This is a logical fallacy known as Argument from Silence (argumentum ex silentio) whereby an argument is deemed valid due to the absence of a theory or evidence to the contrary. Also Argument from Ignorance, whereby absence of another explanation is used to validate a premise.

3. Historical science is the study and attempted reconstruction of events that occurred in the past, as contrasted by science that studies existing phenomena.

4. Panchronic – throughout time.

5. Most of the other 98.5 percent is called "junk DNA" and is only now beginning to yield its functions. Just because scientists have "read" the entire sequence in man and some other creatures, that doesn't mean they understand it. It would be like unearthing a large library of books written in an extinct language that no one understands and declaring that the contents have actually been understood.

6. Richard Dawkins defines biology as "... the study of complicated things that give the appearance of being designed."

7. Teleological definition: (Gk tele or telos = end, purpose) 1a: the study of evidences of design in nature; a doctrine that ends are immanent

in nature; 1c: a doctrine explaining phenomena by final causes; 2: **the fact or character attributed to nature or natural processes of being directed toward an end or shaped by a purpose**; 3: use of design or purpose as an explanation of natural phenomena.

8. "But, beloved, be not ignorant of this one thing, that one day is with the Lord as a thousand years, and a thousand years as one day," 2 Peter 3:8.

CHAPTER 10

1. *Hitchhikers Guide to the Galaxy*, Douglas Adams, 1979. A satire in which instantaneous intergalactic travel is possible due to an infinite improbability drive.

2. Note that the hypothetical numbers given here of amino acids, proteins and DNA nucleosides in a simple bacterium are simplified to make calculations easier to follow.

3. Purines Adenine (A) and Guanine (G) must pair with pyrimidines Cytosine (C) and Thymine (T), only as A-T and G-C to form each nucleotide pair that forms each "rung" of DNA. RNA substitutes uricil for thymine.

4. Fred Hoyle, "Hoyle on evolution," *Nature*, Vol. 294, No. 5837 (November 12, 1981), p. 105

5. Fred Hoyle and N. Chandra Wickramasinghe, *Evolution from Space* (London: J.M. Dent & Sons, 1981)

6. Sigma Life Science, part of Sigma Aldrich Company, St. Louis, MO, USA. www.sigmaaldrich.com

7. "Life," from Wikipedia, https://en.wikipedia.org/wiki/Life, reference 1. Koshland, Jr., Daniel E. (22 March 2002). "The Seven Pillars of Life." *Science* 295 (5563): 2215–2216. doi:10.1126/science.1068489. PMID 11910092.

8. "Tiny Frozen Microbe May Hold Clues To Extraterrestrial Life," *Science Daily* (June 15, 2009) — "A novel bacterium -- trapped more than three kilometres under glacial ice in Greenland for over 120,000 years... Dr Jennifer Loveland-Curtze and a team of scientists from Pennsylvania State University report finding the novel microbe, which they have called *Herminiimonas glaciei*, in the current issue of the *International Journal of Systematic and Evolutionary Microbiology*.

The team showed great patience in coaxing the dormant microbe back to life; first incubating their samples at 2°C for seven months and then at 5°C for a further four and a half months, after which colonies of very small purple-brown bacteria were seen.... and it has been shown that ultramicrobacteria are dominant in many soil and marine environments."

9. Extremophile – microbes that thrive in extreme conditions that would kill other organisms. They have been found in boiling hot water, under extreme pressure, at high altitudes, in sulfuric acid rich waters, in oil wells, et cetera. Almost no place on earth is devoid of life. It is ubiquitous.

10. Table salt is produced in two ways, mines or evaporation of salt water, so it is uncertain if this was an ancient organism. Ponds used to evaporate sea water are often tinged purple or red by halobacteria and must be purified before sale for food products, so salt with dormant microbes was probably mined from deep underground.

CHAPTER 11

1. A coenzyme is a smaller molecule that acts with an enzyme to perform a particular task.

2. GTP (guanidine triphosphate) is the high energy unit in this step (GDP and GMP are the lower-energy units).

CHAPTER 12

1. Robert Shapiro, *Planetary Dreams*, John Wiley and Sons, Inc., 1999.

2. Miller-Urey experiment or simply Miller experiment by Stanley L. Miller directed by Harold C. Urey in 1953.

CHAPTER 13

1. Light year is the distance that light can travel in a vacuum in one year.

2. Solutions by Georges Lemaitre and Alexander Friedman. Einstein's own calculations included a cosmological constant that resulted in a static, non-expanding, universe, which did not fit with the desired progressive picture of others. At one point he supposedly renounced the Cosmological Constant when he told George Gamow that "... the introduction of the cosmological term was the biggest blunder of

his life," although some others who knew him contended that it was a joke. Note that this so-called Einstein quote was only related by Gamow in 1970, not directly from Einstein who died in 1955. Long after rejecting Einstein's cosmological constant, cosmologists have included a new Cosmological Constant, attributed to dark energy, to explain an apparent acceleration of expansion.

3. *American Scientist*, Vol. 30 April 1942 No. 2.

4. Plasma is a completely ionized gas, sometimes referred to as the fourth state of matter along with solid, liquid, and gas.

5. "A Relationship Between Distance and Radial Velocity Among Extra-Galactic Nebulae," Edwin Hubble, *Proceedings of the National Academy of Science*, Vol. 15, 168, 1929.

6. Km/sec/MPc is the recessional speed in kilometers per second, per megaparsec distance. A parsec is the distance to an object that has an apparent motion (parallax) of one second of arc, or approximately 3.26 light years, so a megaparsec is 3.26 million light years.

7. Distance to Andromeda is calculated from Cepheid variable stars, not from redshift. See following text for explanation.

8. Cepheid variable stars with the same period of variability are thought to be the same brightness everywhere. So, distance can be calibrated depending upon their loss of brightness with distance. This was the original Standard Candle and has been joined by a succession of others needed to calibrate greater and greater distances, sometimes with little or no overlap.

9. "History of the 2.7 K Temperature Prior to Penzias and Wilson" A. K. T. Assis, Instituto de Física "Gleb Wataghin" Universidade Estadual deCampinas 13083-970 Campinas, São Paulo, Brasil M. C. D. Neves Departamento de Física Universidade Estadual de Maringá 87020-900 Maringá, PR, Brasil.

10. Newtonian geometry was later modified by Einstein's theory of relativity, but it is now known, ie, believed, that the universe is so nearly Newtonian that in most cases the corrections are negligible. Newtonian geometry deals with flat planes, whereas the Riemannian geometry used in relativity and cosmology uses positively curved (elliptical) planes. The easiest way to think of this is to assume the surface of the earth or other sphere as the plane. (Lobachevskian

geometry has a negative curvature with hyperbolic planes, sort of like a saddle.) So, if mass curves space, and the overall curvature of the universe is nearly Newtonian, the actual curvature must be extremely slight.

11. From very short, highly energetic wavelengths to very long lower energy wavelengths the spectrum includes gamma rays, X-rays, ultraviolet light, visible light, infrared (usually associated with heat), microwaves and radio waves

12. Wavelength (λ) and frequency (ν) are inversely related by the formula $\nu=c/\lambda$ so that the shorter the wavelength, the higher the frequency. Wavelength is usually reported in waves per centimeter (or meter). Frequency is usually reported in waves per second.

13. Exothermic – from exo meaning out and therm meaning heat. Exothermic reactions give off heat, but endothermic reactions absorb heat from their surroundings and become colder.

14. Like Big Bang, Tired Light was a derogatory term used by the theory's opponents.

15. "On the Redshift of Spectral Lines Through Interstellar Space" by F. Zwicky, California Institute of Technology, August 26, 1929, *Proceedings of the National Academy of Science* 15:773-779.

16. Opinion: I have my doubts about this one. Rather than bending of light by gravity, could the bending of light by the sun's extended atmosphere be responsible for most of the observed effect?

17. "The Problem of the Expanding Universe," Edwin Hubble, *American Scientist*, April 30, 1942.

18. "On Supernovae," Baade, W.; Zwicky, F. (1934), *Proceedings of the National Academy of Science* 20 (5): 254–259.

19. "On Collapsed Neutron Stars", Zwicky, F. (1938), *Astrophysical Journal* 88: 522–525.

20. "On the Formation of Clusters of Nebulae and the Cosmological Time Scale", Zwicky, F. (1939), *Proceedings of the National Academy of Sciences* 25 (12): 604–609.

21. "Nebulae as Gravitational Lenses," F. Zwicky, *Phys. Rev.* 51, 290 (1937).

22. Zwicky, F.; Herzog, E.; Wild, P. (1961), *Catalogue of Galaxies and of Clusters of Galaxies* 1,California Institute of Technology, Bibcode:1961cgcg.book and later editions.

23. Wikipedia: article "Fritz Zwicky," Sept. 2010 version.

24. "The Case Against Cosmology," Published in *General Relativity and Gravitation*, Vol. 32, Issue 6, p. 1,125, 2,000 arXiv:astro-ph/0009020v1 1 Sep 2000.

25. Surface of last scattering refers to the belief that the earliest period after the Big Bang event was opaque to light because it was made up of a plasma of bare subatomic particles, ie, electrons, protons and neutrons, and only became transparent to light when atoms were formed.

26. K denotes Kelvin temperature scale. The Kelvin scale is thought to begin at zero for Absolute Zero, the theoretical state in which all molecular motion ceases, ie, heat is zero. For some perspective, $0°$ K = $-273.16°$ C or $-459.69°$ F.

27. Ref: "History of the 2.7 K Temperature Prior to Penzias and Wilson" A. K. T. Assis, Instituto de Física "Gleb Wataghin" Universidade Estadual deCampinas 13083-970 Campinas, São Paulo, Brasil M. C. D. Neves Departamento de Física Universidade Estadual de Maringá 87020-900 Maringá, PR, Brasil.

28. a priori means from the beginning, without precedent.

29. Public Domain image and text sourced from Wikipedia. Original source and its integrity are not known, so this should only be used as a relative power illustration, not absolute and confirmed numbers. The only Wikipedia reference is: Hauser M. G., Dwek E., 2001, *Annual Review of Astronomy and Astrophysics*, 39, 249. (Article titled "The Cosmic Infrared Background: Measurements and Implications").

30. An estimated trillion neutrinos per square meter per second pass through Earth.

31. "From Wikipedia article Neutrino, section: Alteration of nuclear decay rate.

A Russian study suggests that the decay rate of radioactive isotopes is not constant as is commonly believed; [a Stanford study] also finds this, and says it appears to be affected by the rate of neutrinos emitted by the Sun. If true, this finding casts doubt on the absolute reliability of "Radiometric dating" if the neutrino flux from the Sun has not been constant throughout history." Present in the 2011 version, this entire section has been removed from the latest version. References given are

1. S.E. Shnoll, K.I. Zenchenko, I.I. Berulis, N.V. Udaltsova, I.A. Rubinstein (2004). "Fine structure of histograms of alpha-activity measurements depends on direction of alpha particles flow and the Earth rotation: experiments with collimators." http://arxiv.org/abs/physics/0412007

2. D. Stober (August 2010). "The strange case of solar flares and radioactive elements." Stanford Report http://news.stanford.edu/news/2010/august/sun-082310.html.

32. Recently, an excess of neutrinos has been detected at Antarctic IceCube kilometer sized neutrino detectors and has been assumed to be interstellar or intergalactic neutrinos. Source: http://arXiv:1304.5356v1 [astro-ph.HE] 19 Apr 2013.

33. A series of Standard Candles are used to calibrate redshift to distance. From nearest to farthest, they are: Radar, Stellar Parallax, Spectroscopic "Parallax" (really fitting stars into a brightness and color vs. size relationship), Cepheid variables, Tully-Fisher galaxy brightness relationship, Type 1a Supernova, and Redshift alone.

34. Quintessence is a name borrowed from the ancient Greeks who thought the aether was the fifth element (quinta essensia in Latin). Quintessence is thought by some cosmologists to be the fifth fundamental force along with gravity, electromagnetism, strong nuclear force and weak nuclear force.

35. Interpretation of the Casimir effect that supposedly detected quantized vacuum energy between two metal plates is still debated. The original experiment did not involve vacuum energy but has been interpreted as such by others.

36. Entropy is a measure of the disorder of a closed system and the Second Law of Thermodynamics states that entropy always increases – that chaos always increases and usable energy always decreases.

37. Most of these theories claim dark matter fluctuations occurred first and attracted matter.

38. Estimated from the average of 1 hydrogen atom per 4 cubic meters of space assumption mentioned earlier that includes all the matter in galaxies averaged.

39. Blackbody refers to an ideal surface that absorbs all photons. Blackbody radiation is the heat given off by such a surface that depends only on its temperature. Its spectrum is a featureless curve.

CHAPTER 14

1. Manifest as used here means readily perceived by the senses.
2. Quanta is plural for quantum (from the Latin).
3. A standing wave is a wave that is stabilized by resonance or other means. For light or other electromagnetic waves such as radio or microwave, they are stabilized by electric and magnetic fields perpendicular to each other and to the direction of travel. For traveling, non-standing waves, the waves will not persist. For example, for sound in air, the traveling waves are damped and dissipated by the movements and resistance of air. Sound waves traveling in metal railroad rails will persist longer because of reduced interference, but will also gradually die away.
4. Wavelength is the length of a wave from crest to crest in centimeters. In the case of light, measurements are in billionths of centimeters, ie, nanometers.
5. This can also be written as $W = hc\lambda^{-1}$ with λ being the wavelength and c being the speed of light because $v = c\lambda^{-1}$ where v is the frequency. Note that the two Greek letters used are v, pronounced nu, and λ, pronounced lambda, which are roughly equivalent to n and l in our alphabet.
6. An alpha particle is a bare helium nucleus of two protons and two neutrons that is emitted in alpha decay of certain radioactive elements.
7. Momenta is plural of momentum, singular.
8. Schrodinger used partial differential equations in sets of three for the three dimensions of height, width and depth to describe electromagnetic waves that have components in all three dimensions, ie, electronic, magnetic, and direction of travel.
9. It should be noted that many thought experiments and most actual experiments have been done using light, not subatomic particles. The results of these actual experiments depend on your interpretation of quantum theory. See other interpretations that follow.
10. In discussions between Einstein and Neils Bohr about nonlocality/superposition. See next endnote.
11. "Can Quantum-Mechanical Description of Physical Reality Be Considered Complete?" Einstein, Podolsky and Rosen, *Physical Review*, May 15, 1935.
12. Listed major and minor interpretations are from Wikipedia and may not be a complete list. See https://en.wikipedia.org/

wiki/Interpretations_of_quantum_mechanics and https://en.wikipedia.org/wiki/Minority_interpretations_of_quantum_mechanics.

13. Assuming that vacuum energy is a real, and not just a mathematical or theoretical creation.

CHAPTER 15

1. CAFE Standards are Corporate Average Fuel Economy Standards first mandated by Congress in 1975 during the energy crisis. These standards have been continually tightened for even greater fuel efficiency. The result is lighter, smaller, less protective and less safe cars that are contributing to highway crash deaths.
2. Seas have been rising since the Little Ice Age in the seventeenth and eighteenth centuries at about 7 inches per century.
3. Raw data was actually destroyed to prevent others from getting it through a Freedom of Information request.
4. http://wattsupwiththat.com/2010/03/10/when-the-ipcc-disappeared-the-medieval-warm-period/ Article by WUWT guest author Frank Lasner.
5. Recently there have been efforts to define it as climate disruption or climate catastrophe, implying that a tipping point is near.
6. Arthur B. Robinson, Noah E. Robinson and Willie Soon, Oregon Institute of Science and Medicine, 2251 Dick George Road, Cave Junction, Oregon 97523 at http://www.petitionproject.org/review_article.php as a Summary of Peer Reviewed Research.
7. Global Lower Troposphere Temperature (TLT) anomaly, NOAA Advanced Microwave Sounding Unit (AMSU) satellite data through RSS (Remote Sensing Systems) http://www.remss.com
8. Plass, G. N. (1956), "The Carbon Dioxide Theory of Climatic Change." Tellus, 8: 140–154. doi: 10.1111/j.2153-3490.1956.tb01206.x.
9. Like most materials on Wikipedia, they are licensed under the GNU Free Documentation License. Many of these are also part of the Global Warming Art project and additionally available under the Creative Commons Attribution-NonCommercial-ShareAlike License.
10. Transmission and Absorption are inversely related by the formula A = 1/log T.
11. The horizontal axis is a log scale in microns so that the 1 to 10 range is in units of 1 and the 10 to 70 range is in tens.

12. Ppbv stands for parts per billion by volume.

13. Ppmv stands for parts per million by volume.

14. 1 bar is approximately 1 atmosphere of pressure at sea level.

15. Example: Phillips, R. P., Finzi, A.C. and Bernhardt, E. S., 2011. "Enhanced root exudation induces microbial feedbacks to N cycling in a pine forest under long-term CO_2 fumigation." *Ecology Letters* **14**: 187-194.

16. See review article of research papers: "Responses of agricultural crops to free-air CO_2 enrichment" Kimball, B.A., Kobayashi, K. and Bindi, M., *Advances in Agronomy* **77**: 293-368 2002.

17. Iodo-compounds contain iodine derived from seaweed.

18. Source: U.S. National Oceanic and Atmospheric Administration (NOAA): GISP 2 and EPICA Dome C through: http://www.climate4you.com/images/GISP2%20TemperatureSince10700%20BP%20with%20CO2%20from%20EPICA%20DomeC.gif

19. Ross McKitrick, Professor of Economics and CBE Chair in Sustainable Commerce, University of Guelph, Guelph, Ontario, Canada www.rossmckitrick.com

20. Reference: "Nonlinear trends and multiyear cycles in sea level records," S. Jevrejeva, A. Grinsted, J. C. Moore and S. Holgate, *J. Geophysical Research, Oceans*, 12 Sep 2006, DOI: 10.1029/2005JC003229.

21. Quote: "More realistically, ice-shelf deterioration is likely to be a rather slow process, and even for a major and sustained warming trend ice-sheet collapse would take several hundred years, with most of the associated rise in sea level occurring during the final century." From *Nature* 277, 355 - 358 (01 February 1979); doi:10.1038/277355a0 "Effect of climatic warming on the West Antarctic ice sheet," Robert H. Thomas 1.), Timothy J. O. Sanderson 2.) & Keith E. Rose 3.). 1.) Institute for Quaternary Studies, University of Maine at Orono, Orono, Maine 04469, 2.) British Antarctic Survey, Cambridge, UK, 3.)Scott Polar Research Institute, Cambridge, UK Present address: Department of Geophysics, Royal School of Mines, Imperial College, London SW7, UK http://www.nature.com/nature/journal/v277/n5695/abs/277355a0.html

22. Reference: "Extracting a Climate Signal from 169 Glacier Records," J. Oerlemans, Institute for Marine and Atmospheric Research, Utrecht

University, Princetonplein 5, 3584 CC Utrecht, Netherlands. *Science* 29 April 2005: Vol. 308, 675-677, doi: 10.1126/science.1107046.

23. "Global integrated drought monitoring and prediction system," Aengchao Hao, Amir AghaKouchak, Navid Nakhjiri, Alireza Farahmand, *Nature*, Scientific Data 1, Article number 140001 (2914), doi:10,1038/sdata.2014.1

24. "Marine Photosynthesis and Oceanic pH," Sherwood, Keith and Craig Idso, CO_2 *Science* (online journal), Volume 10, Number 34: 22 August 2007 http://www.co2science.org/articles/V10/N34/EDIT.php Center for the Study of Carbon Dioxide and Global Change.

25. See Petition Project at http://www.petitionproject.org/.

26. COWDUNG stands for the "conventional wisdom of the dominant group."

27. Svensmark H, Friis-Christensen E (1997) "Variation of cosmic ray flux and global cloud coverage – A missing link in solar-climate relationships." *J Atmospheric and Solar-Terrestrial Phys* 59(11), July 1997, p. 1225–1232. Also Figure 2 in "Reply to Lockwood and Frohlich – persistent role of the sun in climate change," by Henrik Svensmark and Eigil Friis-Christensen, Danish National Space Center, March 2007.

28. Review Article: "Environmental effects of increased atmospheric carbon dioxide,"

29. Watts Barr 1 in 1996.

CHAPTER 16

1. The probability of forming a single protein enzyme 200 units long from the 20 left-handed amino acids, readily available, is 1 in 20^{200} or 10^{260}, (That's a 1 with 260 zeros after it), which would require 231.4 x 10^{180} attempts each second since the beginning of the universe, all in the same microscopic area. The probability of forming 3000 specific, different enzymes required for the simplest bacterium is 1 in $(20^{200})^{3000}$, a practical impossibility.

APPENDIX A

1. Darwin confused the Cambrian with the "lowest Silurian strata" which was really 100 million years beyond the Cambrian. By the

sixth edition he had corrected this error. At the time there was a controversy about whether the Cambrian was a separate strata or was a part of the Silurian.

APPENDIX B

1. Source: Wikimedia Commons at http://commons.wikimedia.org/wiki/File:Glycolysis_overview.svg.

GLOSSARY

1. *Webster's New Collegiate Dictionary*, Seventh Edition, 1963, used for most of the definitions except for technical or newer terms, for which the appropriate dictionaries and glossaries were used.

INDEX

C

D

G

H

I

indeterminate **178, 179, 232**
Inductive Reasoning **48, 102**
Inflationary Period **145, 149, 158**
Information Paradox **167**
Intelligent Design **29, 54, 60, 117, 118, 119, 294, 295**
IPCC **31, 190, 196, 211, 214, 217, 225, 284, 303**
Irving, Washington **301**

K

Kepler, Johannes **18, 38**
Kuhn, Thomas S **35, 299**

L

labor unions **100**
Larsen Ice Shelf **211**
Lemaitre, Georges **149, 313**
Little Ice Age **31, 55, 189, 190, 191, 192, 207, 211, 212, 218, 219, 227, 319**
logarithm **22, 284**
Low Frequency Array **161**
Luther, Martin **38, 97**
Lyell, Charles **104, 107, 110, 243, 299**

M

MACHO **71, 161, 164, 285**
Magical Thinking **61, 63, 64, 98, 176, 231, 235, 237, 285**
magic word **61**
Malthusian **104**
Malthus, Thomas **99, 104, 246, 247, 249, 258, 294**
Mark Twain **17, 19**
Marx, Karl **100, 103, 299**
materialism **6, 28, 98, 100, 103, 280, 284, 285, 303, 309**
materialist **6, 9, 69, 98, 235**
Maunder Minimum **219**
McKitrick, Ross **209, 320**
Medieval Warm Period **31, 189, 190, 207, 216, 218**
metalicity **165, 166, 167**
metaphysics **16, 279, 286, 302**
meteorite **105, 114**

Planck, Max **175, 182**

Plass, Gilbert **195, 297, 319**

plate tectonics **24, 105, 279, 288**

Plato **98**

Plymouth Colony **99, 100**

Popper, Karl **178, 181, 299, 304**

positron **80, 81, 82, 83**

probability **50, 58, 59, 60, 119, 124, 125, 174, 175, 180, 231, 286, 288, 298, 299 321**

Progressive **1, 9, 95, 98, 102, 103, 111, 117, 138, 150, 234, 235, 237, 263, 313**

progressive nitrogen limitation **200, 214**

Progressivism **5, 98, 101, 102, 103, 104, 118, 234, 235**

prokaryotes **91, 94**

proletariat **100, 235, 309**

proton **80, 81, 82, 102, 283, 307**

Proxima Centauri **71**

proxy data **31, 33, 190, 200, 288**

Ptolemaic **18, 36**

Ptolemy **18**

Q

quantum **24, 25, 81, 156, 162, 167, 169, 173, 174, 175, 176, 177, 178, 182, 183, 231, 287, 318, 319**

quintessence **162, 288, 317**

R

redistribution **226**

redshift **40, 41, 42, 43, 144, 146, 147, 148, 149, 150, 153, 154, 155, 156, 162, 169, 232, 289, 290, 314, 315, 317**

reionized **145**

resonance **174, 318**

RNA **91, 115, 117, 126, 129, 130, 132, 133, 134, 135, 137, 139, 140, 289, 312**

Rosen, Joe **296, 301, 318**

Rousseau, Jean Jacques **99, 109**

Rutherford, Ernest **175**